WHY THE SKY IS BLUE

Götz Hoeppe

Discovering the Color of Life

Translated with John Stewart

PRINCETON UNIVERSITY PRESS   PRINCETON AND OXFORD

LIBRARY OF CONGRESS CATALOGING-IN-PUBLICATION DATA

Hoeppe, Götz, 1969–
  [Blau, die Farbe des Himmels. English]
  Why the sky is blue : discovering the color of life / Götz Hoeppe ; translated with John Stewart.—1st ed.
      p. cm.
  Includes bibliographical references and index.
  ISBN-13: 978-0-691-12453-7 (hardcover : alk. paper)
  ISBN-10: 0-691-12453-1 (hardcover : alk. paper)
  1. Color—History. 2. Color—Philosophy. 3. Optics—History.
  4. Light—Scattering. 5. Atmospheric physics. I. Title.
  QC494.7.H6413 2007
  551.56'6—dc22   2006026520
                  Rev.

British Library Cataloging-in-Publication Data is available

For Brigitte and Carl

Contents

Illustrations

FIGURES

Tables

Acknowledgments

This book is a substantially revised translation of *Blau: Die Farbe des Himmels,* which appeared in 1999. It sets out to explore the sky's blue color by tracing historical attempts to explain its origin. On a journey that takes us from ancient myths through Greek philosophy and the development of ideas about optics, the atmosphere, and visual perception to statistical physics, the ozone layer, and the evolution of Earth's atmosphere, I attempt to demonstrate how interdisciplinary approaches have proved fruitful, indeed indispensable, in understanding this phenomenon. Looking at the sky's blueness and wondering what causes it makes us become part of a long history of human curiosity about the world that we live in.

I have accumulated many debts both during the writing of *Blau* and after it was finished. Writing and researching this book consumed more of my life than I could have imagined when I first began. Many friends and colleagues have helped me over the years by providing suggestions, translating Latin and Italian texts, reading parts of the manuscript, revising line drawings and debating issues of history, philosophy, art, and physics with me. I am particularly indebted to Karin Brandmüller, Charlotte Bigg, C. N. Brown, Hans Brücke, Ulrich Buczilowski, Hans-Jörg Deeg, Georg Elwert, Boris Hermann, Joanie Kennedy, Raymond Lee, Astrid Lindenlauf, Ute Luig, Ralf Lützelschwab, Stefanie Märklin, Konrad Mauersberger, Carmen Müllerthann, Susanne Nickel, Tina Otten, Eveline Passet, Raimund Petschner, Axel M. Quetz, Lord and Lady Rayleigh, Abdelhamid Sabra, Jürgen Schreiber, Ingo Schwarz, Hon. Guy Strutt, Stefan Vehoff, Guillemette Vincent, Hans-Heinrich Voigt, Bärbel Wehner, and Almut Weiler. I am much indebted to the staff of the Staatsbibliothek (Berlin), Kunstbibliothek (Berlin), the library of the Museum für Ethnologie (Berlin), the Bayerische Staatsbibliothek (Munich), the library of the Deutsches Museum (Munich), the British Library (London), and the Universitätsbibliothek Heidelberg. I am particularly grateful to my editors, Katharina Neuser-von Oettingen at Spektrum Akademischer Verlag and Ingrid Gnerlich at Princeton University Press, for their enthusiasm, help, and advice in seeing this project through. Eileen Reeves kindly reviewed the manuscript and provided a number of useful comments. I thank Marjorie Pannell for her painstaking copyediting and Terri O'Prey for her dedicated work in producing this book. I thank Judy Benson for her translation of Chapter 9 and John Stewart for translating and revising most of the remaining text.

Stefanie Märklin has endured much during the many months of writing. I am deeply grateful for her love, patience, and understanding.

This book is dedicated to my parents, Brigitte and Carl Hoeppe.

Why the Sky Is Blue

Looking at the Sky

Is the azure of the sky its true colour? Or is it that the
distance into which we are looking is infinite?[1]

—*Chuang Tzu (fourth century B.C.)*

When I was six years old, my father gave me a copy of *Kosmos,* a popular German science magazine, that had a map of the stars and constellations on a deep blue sky. One night we went out to identify stars and constellations and watch the crescent moon through binoculars. The beauty of these distant worlds was spellbinding. Whenever the sky was clear and my parents allowed me to stay up late, I would continue my observations. Equipped with my father's binoculars, I observed the moon, the planets, the Milky Way, and the fuzzy patches of distant nebulae.

But since starry nights are not too common in Central Europe, I became interested in the daytime sky as well. There was so much waiting to be discovered in it. I noticed the many forms and colors of the clouds, rainbows, ice halos, and sunsets. Compared with the regularity of the night sky, the daytime sky undergoes dramatic changes: clouds change shape as they float by, rain falls, fog rises, and the moving sun casts many different kinds of light into the atmosphere. However, the backdrop to this celestial stage seems always to remain the same: the blue of the sky. Hidden behind the clouds much of the time, it makes a magnificent appearance whenever visible, and then it is difficult *not* to look at it.

Only much later did I realize that I was not the only one who felt this fascination and wonder about the sky. Throughout the ages, people have been captivated by its beauty, pondered its mysteries, and feared its forceful powers. People of all cultures have found the sky's blueness particularly noteworthy. Take ancient Egypt, for example, where it was intimately related to Amun, the creator of Sky and Earth. In a hymn to the sun we read:

> Be praised, you who rise from the primeval waters!
> The gods rejoice in looking at you.
> Appear within your barque,
> when the sky shines on your side,
> in the color of lapis lazuli.[2]

This association of the cloudless sky with the colors of precious stones was not unusual in the ancient Near East. The Jewish tradition in the Torah commanded the Israelites to decorate their dresses with blue threads. The color blue (*techeleth*) referred to the sapphire, like lapis lazuli a blue gemstone. The Bible elaborates further on the celestial symbolism of sapphires. In the book of Exodus, for example, the sapphire is considered an emblem of the blue sky as God's throne, a theme that is heard again in the New Testament. Both the sapphire and the blue sky were thought to be particularly pristine, and as such served as a metaphor for religious purity. Further east, we encounter similar attitudes. In Hinduism and Tibetan Mahayana Buddhism, blue is both the color of the sky (*gaganavarna* or *akasavarna*) and the color of the spirit and its enlightenment. Further north, the god Tengri, the most powerful spirit in the ancient religion of Mongolia, was himself equated with the blue sky. The reign of Genghis Khan knew rituals praising Tengri, for Genghis justified his claim to power by calling himself the god's rightful heir. Even in the more recent past echoes of the symbolism of the blue sky are found in Siberian shamanism. Similarly, in Africa, the Ewe, a people that originated in what is now Ghana, knew of a sky god, Mavu, who wore a blue robe:

> They know different sky gods, among whom Mavu is superior.
> Invisible, he lives behind the skies. They consider the blue of
> the sky his robe and the clouds his adornment.[3]

European mythology also ascribes powerful forces to the sky. Odin, the god of battles and tempests, was thought to wear a blue cloak.

In contrast, contemporary Western culture has rather neglected the blue sky. Nowadays many people seem to see the blue of the sky more often in advertisements than in nature. How long has it been since you consciously watched the sky become blue? When did you last lean back— in a park, on a beach, or in a deck chair—to look at the clear sky? The British art critic John Ruskin demanded that his drawing students do just this, when he suggested they should consider the blue sky a huge natural painting:

> When you look intensely at the clear sky's pure blue you will see that there is a multitude and variety of it in complete repose. It is not just dead color, but rather a profound, vibrating and transparent body of penetrating air, wherein you can imagine short, falling spots of deceiving dust.[4]

Ruskin reminds us that our implicit assumption that we know what the sky looks like is an illusion. We do not know the sky. We may look at it, but we do not expect to see anything but a uniform blueness. Thus, we are oblivious to its many shades and hues. If we take Ruskin's advice and look at the sky as if it were a natural painting rather than a natural background, we may rediscover them. We may also discover that the shades of blue noted by Ruskin are not distributed arbitrarily in the sky but follow a regular pattern. The meteorologist Ludwig Friedrich Kämtz described this pattern thus:

> In general, the sky is darkest at the zenith, and further down the blue color gradually loses its depth until it merges with white at the horizon. However, close to the zenith there may be a con- siderable admixture of white, which depends on the presence of condensed vapours, even if discernible clouds are lacking.[5]

While the blue seems darkest and most saturated over our heads— toward the zenith—both these features diminish toward the horizon (Color Plates 2 and 3). If we carefully examine the sky on a clear day, we will notice that this generalization is slightly wrong: the darkest and most saturated patch of blue appears at right angles to the sun. At the same time, the hue and saturation can vary considerably from day to day. Often the entire sky looks turbid and has, at most, a bluish tinge. A saturated deep blue color is more likely after a rainstorm has moved through or in winter, when the air is particularly clear of dust or "discernible clouds," as Kämtz might have remarked.

We all know that blue and the whiteness of clouds are not the only colors of the sky. Twilight often exhibits a remarkable palette of warm hues, particularly yellow, orange, and red, and at night the sky seems to be black. But even this apparent truism is not strictly true. Under favorable circumstances, such as during a full moon, we may notice that the sky does not appear black or gray but blue. Our eyes are generally not sensitive enough to register this feeble hue; however, long-exposure color photographs leave little doubt about its existence (Color Plate 4).

During a hike up a high mountain on a clear day, the sky's color seems to darken progressively: at higher altitudes the sky loses brightness and the blue becomes a blackish blue (Color Plate 5). This effect can be seen even more clearly from a hot air balloon or an airplane. Lacking this experience himself, the Swiss mathematician Leonhard Euler nevertheless correctly predicted in 1760 what could be seen under such circumstances:

> If it were possible to ascend higher and higher over the surface of the Earth, one would see that the sky's brilliant blue becomes ever fainter; finally, it would disappear in the realm of aether, where the sky appears as black as it does at night. This is because in whatever direction one looks, there is only darkness, for there are no rays of light coming from anywhere.[6]

At the beginning of the twentieth century, Swiss geologist Albert Heim took the voyage Euler could only dream about, ascending in a hot-air balloon to an elevation of 6,000 meters:

> At 4000 or 5000 meters above the sea the Earth seems to become thickly veiled in a blue or blue-violet dust. Forests and meadows become alike in color, streets and rivers are still visible, but rooftops are hardly discernible. Then we can see rivers and lakes only if they reflect light. Everything else seems to be painted over with

a grayish-blue color. We have just traversed the blue sky. Over our heads there is black outer space; beneath us, between us and the ground, is the blue sky. Like a veil illuminated by the sun, it now hides the Earth; when seen from the Earth, it had hid the dark outer space.[7]

Heim realized that Earth's atmosphere, home to the blue sky, is nothing more than a thin veil that surrounds our planet and separates us from outer space. This observation, which he made during a dangerous voyage, has become a common experience for many of us. As latter-day explorers, we traverse the sky in an airplane and look at the dark space above. The view out the window of a jet airplane tells us a simple but breathtaking truth: the atmosphere is the mere skin of our planet. It is also a necessary shield that protects us from dangerous energetic radiation stemming from the sun and other celestial bodies.

The belief that the blue sky is a protective shield for life on Earth is not new. The legends of the Buriat and Ostiak nomads of Siberia describe the sky as a huge tent enveloping the world. The holes in this tent only become apparent at night when light shines through, the stars. The Pleiades is the name given to a cluster of stars in the constellation Taurus that is visible every year from October through the winter. It is thought to be the largest hole in the imaginary tent, and as such it lets in the cold wind. Thus, the appearance of the Pleiades was connected with the cold winter season, reflecting the nomads' experience.

Another popular explanation for the blue sky was handed down in the Near East. It is the legend of the mountain Kâf, which the Persian encyclopedist Zakarija al-Qazwînî wrote down in the thirteenth century:

The interpreters say: this is a mountain that completely surrounds the terrestrial world. It consists of green emeralds that cause the blue-green color of the sky. Behind [this mountain] live men and creatures known only to God.[8]

According to this legend, the mountain's color was reflected in the sky and so was visible from Earth. The common man imagined Earth to be flat and surrounded by an immense ocean, behind which Kâf rose. While the terrestrial world was inhabited by humans and animals, the golden land beyond Kâf was home to the gods and mythical creatures. One of these creatures was Simurgh, a bird that resided on the mountain and left it only to advise kings and emperors. The fact that an emerald-*green* mountain should make the sky look *blue* points to a peculiar feature of the classical Arabic language: it had only one word for blue and green, the color term *ahdar*.

Tracing the sky's blue color to an imaginary nomad's tent or a mountain stretching around the inhabited world seems meaningful within the conceptual frameworks of the respective cultures. Scientific reasoning, however, has pursued a different path. First, rational ideas were developed about the substance of color and the structure and composition of the sky. Only then was it possible to reason about the origin of blue. The story of how humans began to understand the sky's blueness starts in Greek antiquity. In the fourth century B.C., the philosopher Aristotle developed the first systematic doctrine of color and used his ideas to explain the colors of the rainbow and the setting sun. He also elaborated the concept of a spherical world in which the visible sky has a natural place within the spheres of air and fire. His pupil Theophrastus was the first to come up with a "scientific" explanation of its color.

These early Greek attempts to explain the blue color gave birth to a suspicion about its origin that continues even today: the idea that the blue is due to sunlight illuminating atmospheric air in front of dark space. This suggests that the sky's blue is not a material color but a manifest one that is related to the atmosphere's spatial depth. If this is true, then the blue color is created somewhere between our eyes and the upper limit of atmospheric air that borders on outer space. A history of the blue sky must then be a story of our visual perception, the optical effects of air, its illumination by the sun, and its dark background. This might seem to be a paradox. After all, seen from close up, the air looks colorless and completely transparent. Larger amounts of air are necessary to make it visible and give it the appearance of being colored. This is a blessing for us, for without this property the daytime sun would shine in a black sky:

It is of incomparable benefit for humans that the atmosphere reflects light and is not completely diaphanous. Without this property we would be unable to see those places that are solely visible
due to light reflected from the atmosphere [but not directly illuminated by the sun]. The glaring difference between the blackness
of empty space and the bright rays would exhaust the eye, and
perhaps even destroy its vision.[9]

This vivid description in a physics compendium written in 1825 proved a
reality to astronauts: without the air, the sky appears to be black. Diffuse
daylight, even on cloudy days, is also connected with the blue color. And
when the author of the physics compendium writes about light *reflected*
from the atmosphere, he is already referring to modern explanations of
the blue color. The theories involved in these explanations have grown
from the first simple propositions of the Greeks into elaborate, abstract
constructs that are truly comprehensible to only a small circle of special-
ists. This development is marked by such eminent researchers as the Arab
scientists al-Kindi and al-Haytham, the medieval monks Ristoro d'Arezzo
and Roger Bacon, Leonardo da Vinci, Isaac Newton, Johann Wolfgang
Goethe, Lord Rayleigh, and Albert Einstein, to mention only a few. Even
though all of these scholars made fascinating and influential discoveries
concerning the sky's blueness, they approached the subject from a variety
of distinct vantage points. Studying their work makes one reflect on what
constitutes an explanation, and on how the meaning of explanation has
changed over time. For instance, what constituted a satisfactory argument
to Leonardo da Vinci might not have been one for Albert Einstein.

It has been proposed that instead of asking "Why is the sky blue?"
one should ask, "How is it blue?" Doing so shifts our attention to the diverse frameworks in which an explanation may be formulated. This is
how I shall proceed in this book, focusing on the ways in which the
light and color of the sky have been conceived of by humans throughout
the ages.

Critics have often blamed science for thoroughly disenchanting our
world, yet I am sure that most of the people mentioned above were

captivated by the blue sky's beauty without worrying about disenchanting it themselves. As for myself, I am convinced that learning about the wealth of human imagination concerning nature can add often surprising insights to an apparently mundane phenomenon. And if I am lucky, this book will cause you to see the sky with new eyes and a new sense of wonder.

CHAPTER 1

Of Philosophers and the Color Blue

In late 1996, a discovery was made near Rigillis Street in downtown Athens. While preparing the grounds for a new museum of modern art in the Greek capital, workers unearthed the foundations of an ancient building under a sprawling dirt parking lot. Finding ancient ruins under any part of Athens is no surprise, but this was a special case. No major excavation had ever taken place in the area, yet ancient writings indicated that this was roughly the place where, two and one-half millennia ago, the Lyceum had been located. Known to the ancient Greeks as the *Lykeion*, this was a public garden with military marching grounds, a gymnasium, sanctuaries, and groves. Reportedly there were colonnaded walks as well, where the eminent philosopher Aristotle gathered his students for philosophical teaching and discourse. In early 1997, archaeologists from the Greek Central Archaeological Council concluded that the ruins were indeed part of the Lyceum. The news that Aristotle's school had finally been found traveled around the world.

Although the Lyceum is the one place commonly associated with the name of Aristotle (Figure 1.1), for much of his life he was more of a wanderer, and we shall see that his philosophy benefited from the experiences he gathered "on the road." Aristotle was not a native of Athens. He was born in 384 B.C. in Stagira, a town in northern Greece on the border with Macedonia. His father worked as a physician for Amyntas II, the King of Macedonia, and while this intimates the privileged conditions under which Aristotle was raised, it also foreshadows his precarious position between two neighboring countries that were not always at peace. At the age of seventeen, Aristotle was sent to Athens to become a pupil at the Academy of Plato, a famous philosopher who himself had been a pupil of Socrates. Aristotle remained at the Academy for twenty years and eventually became an assistant to his famous teacher. When Plato died in 347 B.C., Aristotle left Athens and went on to travel and study around the

Figure 1.1 Bust of
Aristotle in the Museo
Nazionale, Rome.
Courtesy of Bildarchiv
Preussischer Kulturbesitz,
Berlin.

Aegean Sea and Asia Minor (modern Turkey), pursuing mostly biological studies. During long stays on the islands, he became keenly interested
in marine life, making some observations himself and gathering others
from fishermen, farmers, and physicians. Then he returned to Macedonia,
where he was called to teach Amyntas' grandson, Prince Alexander, who
would later be known as Alexander the Great. Only in 335 B.C., when
Athens fell under Macedonian rule, did Aristotle return to Athens. He
founded a philosophical school in the Lyceum, which soon emerged as a
competitor to the Academy (Figure 1.2).

A few years after the discovery of the ruins on Rigillis Street, more and
more archaeologists came to view the identification of the ruins with the
Lyceum as premature. The building turned out to be a villa from Athens'
Roman period, dating from the second century A.D., approximately four
centuries after the death of Aristotle. Thus, the colonnaded walks where
he strolled remain as yet undiscovered. However, modern scholars agree
that the Lyceum must have been situated nearby.

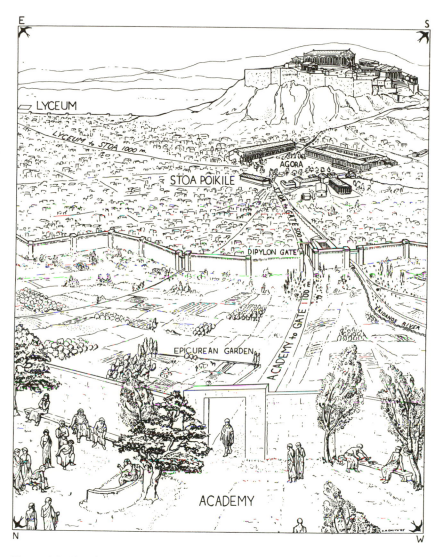

Figure 1.2 Sketch of Athens in the classical Greek period (fourth century B.C.), showing the locations of the important philosophical schools. Courtesy of Candace H. Smith.

Aristotle lived at a time when Greek thinking about nature was changing rapidly. For countless centuries, natural phenomena had been explained solely as the actions of the gods, most notoriously Zeus himself, the father of them all. But starting in the fifth century B.C., scholars at centers of learning began to debate questions such as whether, and how, planetary motion could be understood by means of mathematics, and which substance might fill the space between the stars. Meteorological phenomena drew their attention as well, perhaps because the caprices of weather and the devastation caused by storms were of particular concern to an agrarian society.

In Greek meteorology, two traditions are discernible from that time. One dealt with the prediction of weather, mostly in the form of farmers' rules and almanacs. Throughout antiquity, poetry was an important medium for conveying information, including technical data and instructions, and poems containing rules for predicting the weather were geared toward farmers and sailors as the groups most directly affected by its vicissitudes. The *Phaenomena* of Aratos, a poem of more than a thousand lines, is the most famous example of this genre.[1] Another tradition concerned itself with explanation, attempting to understand the phenomena observed in nature within the framework of a comprehensive theory. There were various schools that followed different approaches and came to different conclusions. Some claimed that the gods did indeed have a marked effect on nature as it was perceptible to man. Others, however, notably the members of the so-called Milesian school, refrained from attributing any perceptible action to deities. This was new. The Milesians assumed that the universe was made of one basic substance, but were unable to agree on what that substance was. Thales surmised that it was water, while Anaximenes claimed that it was air.

Faced with such a variety of conflicting ideas, Aristotle grew uneasy. He envisioned a systematic philosophy that would provide a framework for unambiguously interpreting natural phenomena. Given this background, we must not be surprised that he considered the problem of sensory perception particularly important. After all, our knowledge of nature relies on trust in our perception—whatever we feel, smell, hear, taste, or see. In particular, light and color make the world perceptible to our eyes, vision perhaps being the central mode of perception. Aristotle deals with the senses in several books, especially *On the Soul* (De anima) and *On the Senses and their Perceptions* (De sensu et sensibilia).

As we may infer from Aristotle's devotion to biological studies, natural history was one of his favorite pursuits. And indeed, he was interested in the study of meteorology and weather as well. Once more he realized that contradictory opinions and teachings abounded and that a new doctrine was needed. Yet he also realized that meteorology was a particularly difficult field of study. Meteorological phenomena, Aristotle realized, were halfway between the perfect regularity of the heavens and the irregularity encountered on Earth. Nevertheless, he tried to come to grips with the phenomena observed in the sky and to explain their colors. His book *Meteorology* attests to these efforts.

WERE THE GREEKS BLUE-BLIND?

When we search the works of Greek authors for clues to how they perceived the color of the sky, we are confronted with a mystery. Vacationers are familiar with the intense blue of the Greek sky and the Aegean Sea. It stands to reason that this color would be mentioned, for example, in Homer's *Odyssey*. But that is not the case: a blue sky is mentioned only rarely in Greek literature. Homer, in the third book of the *Odyssey*, describes the ascent of the sun god Helios thus:

> And now the sun, leaving the beauteous water surface,
> sprang up into the brazen heaven
> to give light to the immortals and the mortal men on the earth . . .[2]

Homer's sea is not blue either. When Odysseus was taken captive by the nymph Calypso and thought longingly of his wife Penelope and of his distant home of Ithaca, Homer wrote:

> Him I saved when he was bestriding the keel and all alone,
> for Zeus had struck his swift ship with his bright thunderbolt
> and had shattered it in the midst of the wine-dark sea.[3]

Elsewhere in the *Odyssey* Homer describes the sea as black, white, gray, dark, and purple. Upon further reading, we discover that other Greek authors also write about the sea and the sky without describing them as blue-colored.

Scholars have recognized since the beginning of the nineteenth century that there is a mystery here. Between 1858 and 1877, the British statesman William Gladstone published several articles and books in which he speculated that the ancient Greeks may have had deficient organs for color perception.[4] In other words, he surmised that they were blue-blind. Yet while the color blue is mentioned only rarely in the literature, we have learned since then from archaeological research that, next to yellow, blue was one of the most frequently used colors in Greek painting. The Greeks were not blue-blind.

Careful linguistic study has resolved this riddle by revealing that to ancient Greeks, including Aristotle, luminosity was more important than hue in characterizing color. For example, the Greek words *melas* and *leukos* can be translated not only as "black" and "white" but also as "dark" and "light." The use of the color term *kyanos* is equally ambiguous; it is usually translated as "blue" and gave rise to our modern color term cyan (a green-tinged blue). *Kyanos* referred to a dark color in general and was used to describe emeralds, but it could also manifest itself as a blue, and could even mean black. The difference between black and blue, then, was not so crucial to the Greeks. It was much more important to them that blue bordered on black or dark, and that both of them constituted the dark end of a scale of colors. If we keep this meaning of the word *kyanos* in mind, then much that is found in Greek literature becomes clear. In a book on precious stones, Aristotle's pupil Theophrastus describes the (blue) lapis lazuli as *kyanos*-colored. Homer's *Iliad,* on the other hand, could describe not only the color of steel but also the (probably black) hair of King Priam's son Hector as *kyanos*. A cloak could also be *kyanos*-colored: "a dark-hued veil, than which was no raiment more black."[5]

ILLUMINATING THE "TRANSPARENT"

The oldest-known statements by the Greek philosophers include thoughts about the nature of light (*phos*) and color (*chroma*). In a poem from the early fifth century B.C., the poet Alcmaeon of Croton had already claimed the opposition of light and dark to be the origin of the colors. This assumption, like the color theories of Empedocles and Democritus, remained a speculation in natural philosophy. Empedocles related the four basic colors—black, white, red, and *ochron* (probably a yellow or dull green color)—to the four elements earth, water, air, and fire. Democritus, the originator of Greek atomic theory, interpreted colors as properties of the surfaces of objects and claimed that smooth surfaces appeared white, whereas rough surfaces appeared black.

Plato, a pupil of Socrates, tried to explain the properties of light and color as functions of our perception. He assumed that the eyes emit visual rays that join together with daylight to form a luminous medium. This medium was supposed to convey a material effluence emitted by visible objects into our eyes so that we could perceive these visible things. Plato attempted to reduce sight to a mechanical sensation like touch, explaining the multitude of colors as the result of the different sizes of the particles emitted by the objects. If they were larger than the particles of the visual rays, then the object would appear black; if they were smaller, then we would see white. He described in great detail the combinations that were supposed to produce the multitude of colors we see, and used the rules proposed by Democritus. But in the end, Plato began to have misgivings and gave up, because he feared that attempting to understand colors was tantamount to meddling in the workings of the Demiurge, the creator of the world.

Aristotle did not let such compunctions hold him back when he first began to study the visual process in depth, probably around 340 B.C. In contrast to Plato, he doubted that our eyes could emit visual rays, surmising rather that they passively receive the rays from visible objects; Aristotle was a proponent of the intromission theory of vision. Like the proponents of the extramission theory of vision, he presupposed the existence of a medium that conveys visual impressions, making objects perceptible to our eyes. Aristotle devoted a great deal of attention to this medium, which he called the "transparent" (*diaphanos*). For him, the "transparent" was a property of all media that are more or less translucent. It was that which is visible, only not absolutely and in itself, but owing to the color of something else. This character is shared by air, water, and many solid objects; it is not qua water or air that water or air is transparent, but because the same nature belongs to these two as to the everlasting upper firmament.[6]

Aristotle explained that light would come about if fire (one of the four elements) were found in a transparent medium (like air or water). In contrast, darkness would mean the absence of fire. Colors would then only be visible when acting on a transparent medium. Only in the light would there be color, because darkness is colorless. Aristotle claimed that color is a property of all objects and of the medium that surrounds them, but not of light. This corresponds to our everyday experiences with vision, because we think we see colors on the surfaces of the objects around us. Interestingly, Aristotle did not see light as propagating itself with a certain finite speed. Rather, he imagined that a luminous body changes the entire transparent medium from a state of potential transparency into a state of actual transparency in no time, the perception of light reaching the observer's eye instantaneously.

But how do colors arise? In his treatise *On the Senses*, Aristotle names three possibilities. First, the surroundings can produce a color impression in the eye by means of tiny black and white particles:

> One possibility is that white and black particles alternate in such a way that while each by itself is invisible because of its smallness, the compound of the two is visible. This cannot appear either as white or as black; but since it must have some color, and cannot have either of these, it must evidently be some kind of mixture, i.e. some other kind of color. It is thus possible to believe that there are more colors than just white and black, and that their number is due to the proportion of their components.[7]

The colors can be arranged on a scale ranging from light to dark according to their differing proportions of black and white. Relatively dark colors, like blue, contain mostly black and only a little white, while for lighter colors, like yellow and orange, the proportion of white predominates (Figure 1.3). This derivation of all colors from the complementary

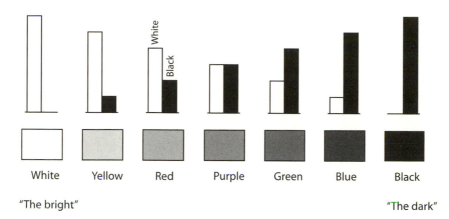

White   Yellow   Red   Purple   Green   Blue   Black

"The bright"   "The dark"

Figure 1.3 The colors produced by black and white mixed in different proportions, according to Aristotle's treatise *On the Senses.*

pair black–white exemplifies a recurrent theme of Aristotelian natural philosophy: pairs of opposites like hot and cold, wet and dry, or light and heavy are commonly employed in ordering phenomena. Aristotle names five colors between white and black: yellow, red, purple, green, and blue. The proportion of white particles decreases in this order; the lightness of these colors therefore also decreases. Blue is next to black and contains very little white. Aristotle thus lists a total of seven basic colors, including white and black. He claims that especially pure colors result when black and white are mixed in proportions of low whole numbers (such as 3:2 or 4:3). Here he is alluding to Pythagoras, who had deduced octaves and the other pure intervals in music from the harmonic relations of tone pitches. This parallel is not coincidental, writes Aristotle, because colors are indeed related to tones.

For Aristotle, the second way that colors can be produced is through the physical permeation of colored substances, from which a mixed color emerges.

New colors may also appear when a luminous color shines through a transparent medium. In *On the Senses,* Aristotle writes:

Another theory is that they appear through one another, as sometimes painters produce them, when they lay a color over another more vivid one, e.g. when they want to make a thing show through water or mist; just as the sun appears white when seen directly, but red when seen through fog and smoke.[8]

This is the third way in which colors can be produced. The effect of smoke and fog is one example; it shows how the air can change the color impression of the sun shining through it. Aristotle identifies a situation in the sky where this transmission of light through a medium has a perceptible effect—namely, the red color of the sun. Could a similar explanation be found for the blue of the sky?

THE SPHERES OF AIR AND FIRE

In searching for an answer, we must consider what "sky" may have meant to the Greek philosopher. More than a century prior to Aristotle's thoughts on light and color, Empedocles had proposed that, in addition to the elements earth, water, and fire, the world was also made up of a fourth element, air. In his book *On the Heavens* (De caelo), Aristotle modifies this idea and divides the world into the terrestrial regions, which he places inside the orbit of the moon around Earth, and the celestial spheres, which are situated outside the lunar orbit. In his view, earth, water, air, and fire are the elements of the terrestrial region. These are characterized by their tendency to occupy certain spaces in accordance with their characteristic natural motions. The heavy elements, earth and water, have a tendency to sink down. The light elements, air and fire, rise up. For Aristotle, the world is finite and has a center. This center is the natural place for the heaviest element, earth. In accordance with their weight, the other elements nestle around Earth in spherical shells and create the spheres of water, air, and fire (Figure 1.4). The sphere of fire, says Aristotle, extends up to the lunar orbit and marks the upper boundary of the terrestrial region. This is the region where everything that comes to be or passes away runs its course: the four elements can transform into one another there. In contrast, he regards the celestial spheres as immutable and eternal. There the moon, sun, planets, and fixed stars move in circular orbits around Earth. Both the celestial bodies

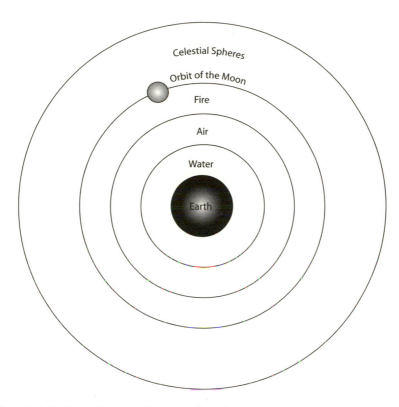

Figure 1.4 In Aristotle's cosmology, the four elements earth, water, air, and fire form concentric spheres around the center of the universe. Outside the lunar orbit, the sun, planets, and stars move in perfect circular paths around this central point.

and the space between them consist of quintessence or aether (*aither*), the perfectly transparent "eternal element of the heavens."

When a body moves toward its natural place, Aristotle considers it an instance of natural motion. The impetus for this motion lies solely in the inner nature (*physis*) of the body. When a body does not move toward its natural place, it is a case of forced motion. The impetus for this motion must be an outside mover. Natural motion has the character of a development and is goal-oriented (teleologic).

One might ask why the celestial spheres move if they consist of the perfect fifth element, aether. In his *Metaphysics*, Aristotle answers that their orbit is special, because it is the only thing that is constant and eternal. Yet they have an originator that does not itself move but remains still. Aristotle calls this being the unmoving mover or the first mover. The ultimate cause of motion must be a living god. This god does not move the celestial spheres by means of a physical push but rather by inspiring in them the desire to approximate his perfection by executing a perfect motion.

While he regarded the celestial spheres and their bodies as the object of study for astronomy, Aristotle located the objects of study for meteorology primarily in the spheres of air and fire—that is, between Earth and the lunar orbit. In doing so, he took the Greek term *meteorologica* literally. It is a combination of the words *meta* (over), *area* (air), and *logos* (the study of). Yet for Aristotle, meteorological phenomena, or *meteora*, included not only rain, evaporation, wind, clouds, and rainbows but also comets, the Milky Way, and even earthquakes and volcanic eruptions. If this list of *meteora* appears to us today to be a hodge-podge of phenomena from meteorology, astronomy, and geology, to Aristotle it was a logical grouping. He considered it the task of meteorology to investigate the exhalations of Earth, as well as how the heat of the sun and the other stars influence them. Transformations between the elements earth, water, air, and fire have a special significance here. Because the spheres of air and fire interact with Earth, or are influenced by objects in the nearer celestial sphere, a systematic study must not ignore them. This is the agenda to which Aristotle dedicated his book *Meteorology*.

Like *On the Heavens* before it, *Meteorology* is part of a series of notes for lectures that Aristotle presumably held in the Lyceum. The lectures were meant to present a comprehensive cross-section of the entire natural world. Aristotle does not assume that there is one physics for the entire universe, but rather that the laws within the celestial spheres and within the terrestrial region are independent of each other. He therefore has to find explanations for the phenomena in the latter that do not assume the perfection of the celestial spheres. Aristotle takes pains to place different meteorological phenomena within a systematic structure, but the imperfection of the terrestrial region entails that this task can only meet with a certain degree of success. Definite knowledge is not guaranteed.

Aristotle bases his remarks on observations and reports of *meteora*. His sources are numerous. He made many observations himself, but also heard second- or third-hand reports of rare events from others. Furthermore, he records earlier attempts at explanations, most of which he then vehemently criticizes. Today we profit from this presentational tactic, because many of the writings of Aristotle's predecessors have been lost. It would otherwise be virtually impossible to reconstruct early classical meteorology.

Aristotle conceived of meteorology as a scientific discipline, but not all of his contemporaries thought likewise. His teacher, Plato, characterized the *meteora* as "lofty things," and left it at that.[9] Although geometry and astronomy had already been developed into systematic theories at that time, some disputed that it was possible to study air, or found the idea of doing so laughable. The dramatist Aristophanes, who belonged to Socrates' generation, wrote in his popular play *The Clouds*:

> Why, for accurate investigation of meteorological phenomena, is it essential to get one's thoughts into a state of, er, suspension by mixing small quantities of them with air—for air, you know, is of very similar physical constitution to thought—at least, to mine. So I could never make any discoveries by looking up from the ground—there is a powerful attractive force between the earth and the moisture contained in thought.[10]

We can take this as evidence that more than a small minority were concerned with meteorological phenomena at that time. Otherwise this speech would not have evoked a comic effect among the audience.

EXHALATIONS

In *Meteorology*, Aristotle explains his ideas on the stratification of the spheres of air and fire, as well as their composition:

> We must understand that of what we call air the part which immediately surrounds the earth is moist and hot because it is

vaporous and contains exhalations from the earth, but that the
part above is hot and dry. For vapour is naturally moist and cold
and exhalation hot and dry: and vapour is potentially like water,
exhalation like fire.[11]

The substance of the spheres of air and fire is the exhalations that rise
when Earth is warmed by the sun. Aristotle differentiates between
two types of exhalations. One is similar to vapor and can be traced
back to the moisture in the earth. It is hot, moist, and heavy. The other
is similar to smoke. It comes from the earth itself and is hot, dry, and
windy.

The ascent of the exhalations is one stage in the cycle of coming-to-
be and passing-away, which is characteristic of the terrestrial region.
There, each of the elements can be transformed into the others: water
becomes air through evaporation, and air becomes water through rain.
These transformations follow laws that result from the composition of
the elements. Aristotle assumes that each of the four elements is a combination of the properties wet or dry and hot or cold. Water is wet and
cold, whereas air is wet and hot. If you want to change water into air,
you have to heat it up. In contrast, air becomes water through cooling
(Figure 1.5).

All these processes clearly happen in what we would today call the atmosphere; they seem to be accurately designated as *meteora*. Other phenomena considered *meteora* by Aristotle may not seem to fit so well into
this category, for instance earthquakes and the Milky Way. To explain the
latter, Aristotle invokes the action of the celestial spheres. Celestial bodies themselves are not hot, but their motion can inflame and rarefy the
air below. This heat adds to that of the friction between the sphere of
fire and the celestial sphere, an effect of their different rotational speeds.
The resulting heat causes the air to glow, and this is visible in the form
of the Milky Way.

Throughout *Meteorology*, Aristotle compares Earth and the cosmos to
the human body. An example is his treatment of earthquakes, which
draws an analogy between winds internal to the body and those inside
Earth, both causing violent motions:

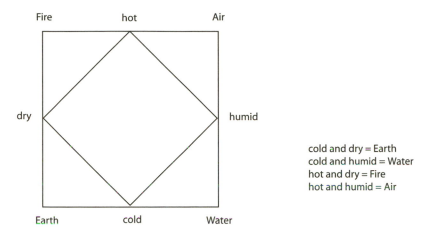

Figure 1.5 Aristotle derived the elements earth, water, air, and fire from combinations of the properties hot or cold and moist or dry.

> We must suppose that the wind in the earth has effects similar to those of the winds in our bodies whose force when it is pent up inside us can cause tremors and throbbings, some earthquakes being like a tremor, some like a throbbing.[12]

Aristotle's use of analogies like this does not mean that he considered Earth or the cosmos to be giant animals, or even to be alive. Instead, it is a sign of his conceding defeat as a scientist: some meteorological phenomena are just too far away or too difficult to understand, so that grasping an aptly chosen analogy is the most one can hope for.

Aristotle is content with describing the macroscopic properties of the four elements and the processes of their transformation; he does not speculate on whether the elements possess specific microscopic structures. This is quite unlike his teacher Plato, who had conceived of the elements as being composed of atoms: tiny, regular, geometric bodies. Plato imagined the atoms of fire to be tetrahedrons, those of water icosahedrons, those of the earth cubes, and those of air octahedrons (Figure 1.6). Aristotle objected to this view, but he agreed with Plato that the elements are inherently colorless.

| Fire | Air | Water | Earth | Kosmos |

Figure 1.6 Plato conceived of the elements as being composed of atoms: tiny, geo-
metrically regular bodies. He imagined the atoms of fire to be tetrahedrons, those
of water icosahedrons, those of earth cubes, and those of air octahedrons. The do-
decahedron, the regular solid closest to the sphere, was associated with the cosmos
as a whole.

FEELING THE AIR

Aristotle was not alone in drawing parallels between meteorological
phenomena and the human body. Throughout Greek antiquity, it was
commonplace to construe the nature of air based on knowledge of one's
own body. This was not necessarily easy, because air is invisible, has no
taste or smell, and we are not able to touch it. All of our experiences
with air are indirect and rest on our perception of it via our own bodies,
when we breathe and feel the wind. In the sixth century B.C., Anaximenes
of Miletus had conceived of air (aer) as the basis of the processes of condensation and rarefaction, from which he derived the development of all
things. The historian Plutarch reported that Anaximenes had carried out
one of the first scientific experiments when he noted that warm air escapes from the relaxed lips of sleepers but cold air comes from pursed
lips. Consequently, Anaximenes considered sleep and relaxation as warm,
but tension and pressure as cold. It followed that there were two types
of air.[13]

Homer had also distinguished between two types of air. He characterized the "clear air aloft under the shining sky" as aither, but "dull air,
fog, haze near the earth, over which a very tall tree up in the aither can
loom" as aer.[14] The distinction between aither and aer was maintained for
a long time. What was meant by these terms, however, changed again
and again, probably most dramatically in the fifth century B.C. Democritus, who lived in the second half of that century, mentioned an old

prayer in which believers lifted their hands to that "which the Greeks now call *aer*."[15] Apparently, he did not wish to suggest that they stretched out their hands to the fog or haze. That is why he emphasized that he meant the current usage of the Greeks; he was simply indicating that the ancient people prayed with their hands in the air—that is, up high.

When Empedocles added air to earth, water, and fire as the fourth element, he still alternated between the terms *aer* and *aither*. Later, he decided to use only the term *aer* for air. *Aither* became the substance of the heavens—and remained so for two millennia.

SEEING THE AIR

Seen from up close, air is clear, colorless, and seemingly fully transparent. But when it condenses, Aristotle writes in *Meteorology*, it can create different colors in the sky:

> [I]t is therefore to be expected that this same air in process of condensation should assume all sorts of colours. For light penetrates more feebly through a thicker medium, and the air when it permits reflection, will produce all sorts of colours, and particularly red and purple: for these colours are usually observed when fire-colour and white are superimposed and combined, as happens for instance in hot weather when the stars at their rising or setting appear red when seen through a smoky medium.[16]

It thus seems that there are two ways in which colors arise in the atmosphere. One is via the attenuation of light, the other via its reflection. Both ways presuppose condensed air, whereas thin air apparently lets the colors of objects pass through unhindered. The idea that a reflective layer of air is supposed to cause "all sorts of colors" alludes to his theory of the colors of the rainbow, which takes up a substantial part of *Meteorology*. Aristotle explains rainbows as an optical phenomenon that results from the reflection of sunlight off a cloud saturated with moisture, and for which the observer must have the sun at his back (Color Plate 6). In his opinion, a rainbow consists of only three colors, namely (from the inner

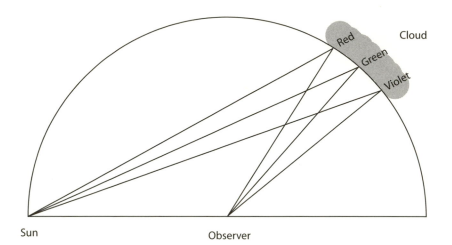

Figure 1.7 Aristotle's explanation of the rainbow.

part of the arc outward) violet, green, and red. Aristotle considers the
yellow that appears in the rainbow to be an optical illusion. Since the sequence of the three colors is the same in every rainbow, he infers a universal principle behind it.

   To explain these three colors, Aristotle first assumes that the individual vapor particles in the cloud work like small mirrors that reflect the
white sunlight. This reflection is one of three causes responsible for creating colors, together with the attenuation of sunlight, as well as the
weakening of vision at great distances. Notice that this contradicts his
claim that there is no such thing as a visual ray. Aristotle goes on to say
that, to a strong eye, reflected colors appear red, to a weaker eye green,
and to an even weaker one violet. According to the color scale described
in *On the Senses*, then, weaker eyes would see everything in the darker
colors, since the degree of lightness decreases over the span of the three
colors from red to violet.

   Departing from the usual style of *Meteorology,* Aristotle makes use of a
drawing to elucidate the relative position of the cloud to the observer
and the sun (Figure 1.7). He places the observer in the center of a semi-
circle and arranges the sun and the cloud along the circumference. He

knows that, as a heavenly body, the sun is much farther from an observer on Earth than a cloud, which he locates in the sphere of air. Thus, instead of showing the distances in their actual proportions, his drawing depicts the apparent dome of the sky, in which both the sun and the cloud are located.

Based on this depiction and on what has been said so far, it is not hard for Aristotle to explain the sequence of colors. We can infer from the illustration that the upper front part of the cloud is somewhat closer to the sun than its middle part and especially its bottom. At the same time, the entire front region is equidistant from the observer. The path of the light rays from the sun by way of reflection off the cloud to the observer is thus shortest when the rays are reflected off the upper edge of the cloud, and longest when they are reflected off its bottom edge. It is exactly the same with the visual rays, whose existence Aristotle surprisingly assumes here, and which are supposed to fan out from the observer to the sun and back. Two factors weaken the light and visual rays: their being reflected, and their traveling some distance. The latter attenuation increases with the distance traversed. Since the color red corresponds to the slightest attenuation of the visual rays, and since the path of the rays by way of reflection off the upper edge of the cloud is the shortest, the upper edge of the rainbow has to appear red. The situation with the other two colors, green and violet, is similar; Aristotle explains them on the grounds of increasing attenuation of the visual rays over their longer paths.

This creative application of visual rays with regard to the rainbow would seem like a way to explain any colors whatsoever in the atmosphere. But Aristotle does not settle for such an easy solution; he distinguishes the colors of the rainbow from those of low-lying stars. Seen through fumes, the latter appear red. The best example of this is the rising and setting sun, whose intense red has always fascinated humankind (Color Plate 7). In *On the Senses,* where he vehemently rejected the hypothesis of visual rays, Aristotle already recognized in this color the effect of something light shining through something dark (Figure 1.8). Thus, with or without visual rays, the condensation of the air is a necessary condition for the emergence of colors in the atmosphere.

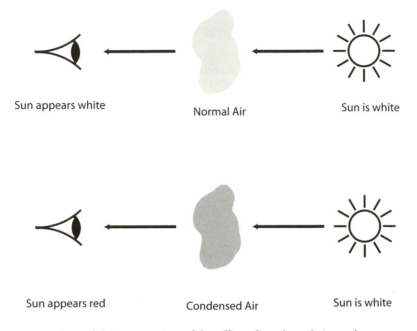

Figure 1.8 Aristotle's interpretation of the effect of condensed air on the apparent color of the sun.

THE PENETRATING DARKNESS OF THE DEPTHS

Aristotle died in the year 322 B.C. Shortly before then he had gone into exile to escape a death sentence. Alexander the Great had died unexpectedly the previous year while on a military campaign. Alexander had subjugated Athens at the beginning of his reign, and so anti-Macedonian sentiments broke out there after his death; these were also directed at Aristotle, as the monarch's former teacher. Aristotle reportedly claimed that he left Athens in order to prevent the city from committing yet another sin against philosophy. Seventy-seven years earlier, the condemned Socrates had been forced to drink a cup of poison.

Even before founding the Lyceum, on his travels in the Aegean, Aristotle had made the acquaintance of Theophrastus of Eresus, who assisted him with his biological research. Theophrastus accompanied Aristotle to

Athens. After Aristotle's death, he was appointed head of the Lyceum and held the position for the next thirty-five years. Theophrastus shared his teacher's interest in meteorology, as can be seen in his work *Metarsiology,* a book concerned with "things in the sky." In the surviving text, Theophrastus distinguishes between phenomena above Earth—to which he assigns thunder, lightning, clouds, rain, snow, hail, dew, frost, wind, and halos—and those below the surface of Earth, which include earthquakes. Theophrastus also dealt with color theory. He is now considered to be the author of the work *On Colors* (De coloribus), which was attributed to his teacher up until the twentieth century.

Even though *On Colors* clearly does rely on the Aristotelian color theory, Theophrastus does not follow his teacher in every detail. Both philosophers agree that the transparent medium is a prerequisite for vision and that all colors result from a mixture of black and white, light and dark. In contrast to Aristotle, Theophrastus places colors in relation to the four elements. Fire (as well as the sun) is golden, air and water are white. Earth is also white, but appears as if "stained" with several colors. While black, for Theophrastus, is a color that can result from some property of a material, it primarily represents the absence of light. He considers sea foam and snow to be proof of the white color of air, interpreting both of them as its compressed states.

We cannot see the colors in their natural purity, writes Theophrastus, because they are always altered by light and shadow. The colors of all objects are influenced, namely, by their illumination, the nature of the medium through which the light rays radiate, and the background they are positioned in front of. Thus the colors of objects appear differently when they are seen in sunlight versus in shadow, in hard or soft light. Media that are dense and transparent appear cloudy, Theophrastus writes, mentioning water, glass, and dense air as examples. The density of these media attenuates the light rays that penetrate them:

> This also happens, one would suppose, in the case of air. So that
> all colours are a mixture of three things, the light, the medium
> through which the light is seen, such as water and air, and thirdly,
> the colours forming the ground, from which the light happens to

be reflected. But the white and the transparent, when it is very
thin, appears misty in colour. But over what is dense a haze invari-
ably appears, as in the case of water, glass and air, when it is dense.
For, as the rays from all directions fail owing to the density, we
cannot see accurately into their inner parts. But the air when ex-
amined from nearby seems to have no colour (for owing to its
thinness it is controlled by the rays and is divided up by them, because they are denser and show right through it), but when exam-
ined from in depth, the air appears from very nearby to be blue
(*kyanos*) in colour because of its rarity. For where the light fails,
there, being penetrated by darkness at this point, it appears blue
(*kyanos*). But when dense, just as with water, it is the whitest of all
things.[17]

Here Theophrastus explains not only the blue appearance of rarefied air.
Imagining clouds to be condensed air, he also ponders their white color,
as well as the blue of the sea. This blue, as the color of light penetrated
by darkness, follows from the color scale described by Aristotle in *On the
Senses:* it is adjacent to black. Theophrastus joins his teacher in declaring
the density of air to be the deciding factor in creating this color. However, his way of accounting for the interactions that lead to the prepon-
derance of darkness is new. The density of the air accumulates primarily
because of its great depth, in which only blue can emerge. This resonates
with the daily observation that air is colorless from close up, and only in
the sky does the blue color appear. At great depth, darkness predominates
because the light rays lose their strength. Theophrastus thus refrains from
assuming visual rays and accounts for the darkness of depth solely with
the attenuation of light rays.

Theophrastus does not explicitly state that he intends to explain the
color of the sky with this argument. Yet there can be little doubt that he
did, because in his worldview air is the element of the spheres of air and
fire, which in turn are seen in the sky. The great depth of these two
spheres, which extend all the way to the lunar orbit, had to suffice to
penetrate the sunlight with their darkness.

CHAPTER 2

A Blue Mixture: Light and Darkness

From the Montparnasse railway station in Paris, you can get to the high Middle Ages within an hour and a half. All you have to do is take a commuter train, walk a bit—and open your eyes. The train leaves Paris' futuristic train station, passes suburbs spattered with immense concrete residential buildings, enters a hilly landscape, winds through forests and fields, and passes villages and farms. Then you reach Chartres, a provincial town 90 kilometers south of the French capital. Before arriving, you have already glimpsed the small town dominated by two towers, pointed arches and a green roof: Chartres Cathedral (Figure 2.1). You can easily reach it on foot from the train station. After entering the cathedral through the west portal—famous for its carved sculptures dating from the late twelfth century—your eyes first need to adapt to the profound darkness of this old church. Then you notice the stained glass above. More than 170 colorful windows provide you with a panorama of the worldview of the high Middle Ages. You discover in them scenes from the story of creation and the great deluge to the life of Jesus, and an overview of medieval arts and crafts. Blue is the windows' dominant color, usually featuring as their background. The most luminous blue of all is near the south side of the altar: the window of the *Blue Virgin*, the *Notre-Dame de la Belle Verrière* (Color Plate 8).

The deep blue of the windows is a trademark of the early Gothic style, and Chartres Cathedral is one of its prime examples, having survived for eight centuries. This colorful glass was first used extensively in the 1140s, when the Abbey Church of Saint Denis, on the northern outskirts of Paris, was restored. Its symbolism continues to be a matter of debate among art historians, but there is little doubt that the meaning of the blue color derives from the clear sky, which was commonly associated with the divine and with sacred purity, their ethereal glow suggesting a mythical dimension of spirituality.

Figure 2.1 Chartres Cathedral as seen from the southeast.

Looking at the Chartres windows more closely gives one a hint as to
how nature was perceived in the late twelfth century. There is a panel
depicting Charlemagne looking at the Milky Way, and there are many illustrations of plants and animals. A number of panels suggest that nature
was understood to be made up of the four classical elements, earth, air,
fire, and water. One panel shows Noah looking at the rainbow after the
deluge has ended, with the bow depicted, from the inner part of the arc
outward, in green, yellow, and red bands. This deviates from the order
that Aristotle had described in his *Meteorology*, which was purple, green,
and red from the inner part of the arc outward. Both series agree with
the actual rainbow, in which yellow appears between green and red.
The discrepancy suggests that the window-makers may have disagreed
with the Greek philosopher, or that they did not know his work. In the
entire pictorial program of the windows, there is no reference to Aristotle's philosophy of nature. We know today that the window-makers

could not have been familiar with it, because throughout most of the Middle Ages, European scholars had been deprived of the Greek philosopher's works.

In A.D. 84, more than a millennium prior to the making of the Chartres windows, the Lyceum, the Aristotelian school in Athens, had been closed. In A.D. 529, the Roman emperor Justinian shut down the city's last philosophical school, and thus began the so-called Dark Ages. When the West Roman Empire fell at the hands of the Huns and Germanic tribes in the sixth century, the works of the classical philosophers were preserved only in the libraries of Constantinople, today's Istanbul, where they faded into oblivion.

FROM ATHENS TO BAGHDAD

The works of classical Greece remained forgotten until the ninth century, when they were rediscovered by Arab scholars. While the founding of Islam by the prophet Mohammed in the early seventh century had been accompanied by numerous wars and military conquests, the political situation in the Near East stabilized during the eighth century. This provided the leisure necessary for the pursuit of philosophy.

The Arabian golden age began in the middle of the eighth century, when the Abbasid dynasty relocated to Baghdad. As descendants of Mohammed's uncle al-Abbas, the members of this dynasty claimed to be the prophet's rightful heirs. The atmosphere at their court was cosmopolitan; they allowed Christian doctors to treat them, and they maintained contacts with Syrian and Persian scholars. There were even important influences from India. Art and culture found a generous patron in Harun al-Rashid, the legendary caliph of the *Arabian Nights*. Among scholars, Greek philosophy enjoyed wide renown, and when Harun learned that manuscripts of Plato's and Aristotle's works were archived in Constantinople, he dispatched ambassadors to search for them. A few months later, a caravan with the precious writings arrived in Baghdad, and the work of translating them into the Arabic language commenced. For this purpose, one of Harun's sons founded the House of Wisdom, a research institute where the Greek works were translated and annotated. The scholars were well aware of the magnitude of the undertaking, and it was to take almost two hundred years to translate virtually the entire corpus of Greek texts about philosophy, medicine, and the natural sciences. We therefore have Arab scholars to thank for the fact that many writings of the Greek philosophers have been preserved up to the present. Arab scientists not only translated the ancient works, they also built on them by conducting research of their own. And some of them contemplated the blue color of the sky.

THE WEAK LUMINOSITY OF THE EARTHLY HAZE

The first important philosopher at the House of Wisdom was Abu Yusuf Ya'qûb Ibn Ishâq al Kindi. Al-Kindi was born around the year 800 in present-day Iraq. After completing his studies in Kufa and Baghdad, he was commissioned by the caliph al-Mamun to revise the Arabic translations of Greek books. Al-Mamun's successor appointed al-Kindi as his son's teacher. Al-Kindi earned high esteem but eventually fell out of favor with the caliph al-Mutawakkil and spent his final years in isolation. He died around 866.

Al-Kindi's writings deal with questions of philosophy, theology, medicine, astronomy, and optics. Full of enthusiasm and reverence, he gushes about the achievements of the classical philosophers. It would be impossible for him and his contemporaries to accomplish anything comparable, he concedes. Rather, he considers it his duty to preserve and restore the old texts, making a few corrections here and there. At the center of al-Kindi's philosophy is the assumption that each thing in nature emits energy rays that fill up the entire world. Viewing the world as a gigantic network whose components are in constant interaction with each other due to the energy they radiate and receive, he considers the study of these rays to be the natural basis of all the sciences. He assumes that visual rays exist, invoking Euclid, who had developed a mathematical theory of the visual process around 300 B.C. Contrary to Aristotle, al-Kindi thus claims that our eyes emit rays that strike visible objects, making these objects perceptible to us.

Among al-Kindi's less known works is a treatise "regarding the cause of the azure-blue color (in Arabic, *lazward*) which is seen in the air in the direction of the sky and which is considered to be the color of the sky." This is the oldest preserved written text devoted solely to this issue. A copy of this work has survived in the Bodleian Library of Oxford University. Its importance was recognized by the orientalist Eilhard Wiedemann, who in 1915 translated it from Arabic into German. An additional copy was discovered a few years later in a library in Istanbul.

In this treatise, the Arab scholar refers to Aristotle's theory of vision and temporarily puts aside the concept of visual rays, of which he was otherwise a vehement proponent. He begins by considering which things can be seen, and whether air belongs to this category of visible things. Al-Kindi claims that only discretely delimited bodies can be illuminated and take on a color. But as an element of the spheres of air and fire, air is incorporeal and amorphous. It is therefore dark by nature and cannot take on any color. This constitutes a problem in explaining the color of the sky, the cause of which al-Kindi suspects lies in the optical effects of air. While Aristotle wrote that air could appear colored when it is condensed, and Theophrastus sidestepped the issue by designating air as white, al-Kindi had to seek a new way to shed light on air. He found it by invoking a phenomenon with which he was well acquainted from the deserts of the Near East. Surely he would have often seen sandstorms that conveyed large quantities of "earthly particles" into the air, making the sky hazy. This haze, al-Kindi writes, extends up to a certain height above the surface of Earth. It consists of fine particles with solid surfaces. This property enables the haze to be illuminated in a passive manner. As a result, the darkness of the air is defeated, and the air is lit up. This light arises from sunrays illuminating the haze, as well as from the heat reflected off the earth:

> The air surrounding the Earth gets weakly lighted by the earthly particles dissolved in it and changed into fiery ones due to the

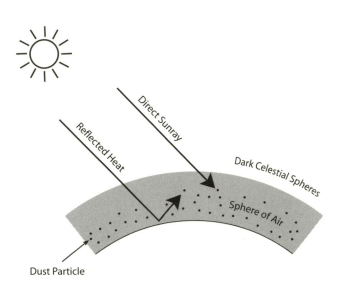

Figure 2.2 Al-Kindi suspected that particles of haze, having been stirred up into the sphere of air, are excited into glowing by the light of the sun, as well as the heat reflected off the surface of Earth.

heat which they have accepted from the reflection of the rays off the earth. The shadowy air above us is visible because the light of the Earth and the light of the stars intermingle into a color in the middle of shadow and light and this is the azure-blue color (*lazward*).[1]

In talking about the transformation of earthly particles into fiery ones, al-Kindi is alluding to Aristotle, who had explained that light was the presence of something fiery in a transparent medium. According to al-Kindi, the "earthly particles" make air glow because they are irradiated by sunlight and by the heat reflected off Earth. The air thus glows with a color that lies "in the middle of shadow and light." This luminosity is weak; it is closer to darkness than to light. The mixture therefore produces a color that borders on darkness: blue (Figure 2.2).

At a crucial juncture, then, al-Kindi has recourse to Aristotle. He explains the blue color according to a formula given by the Greek philosopher a millennium earlier in his work, *On the Senses*.

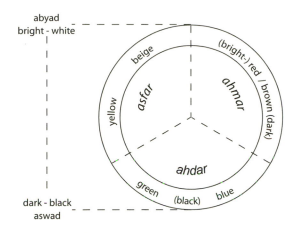

Figure 2.3 Diagram of the color theory found in classical Arabic literature. As in the Greek model, the opposition of black and white formed the basis of the color system. After Wolfdietrich Fischer, *Farb- und Formbezeichnungen in der Sprache der altarabischen Dichtung: Untersuchungen zur Wortbedeutung und zur Wortbildung* (Wiesbaden: Harrassowitz-Verlag, 1965), 237. Courtesy of Harrassowitz Verlag, Wiesbaden.

These explanations show that the Arab philosopher subscribed to the Aristotelian color scale as a sequence ranging from light to dark. Likewise, in the classical Arabic language, the opposition of light and dark determined how colors were named. They were ranked according to their lightness (Figure 2.3) between the poles *abyad* (light, white) and *aswad* (dark, black). In a popular work on colors, the thirteenth-century Persian scholar Nasir al Din al-Tusi described several paths from white to black. This stands in contrast to Aristotle, for whom the proportions of white and black alone determined the color of their mixture. One of the paths proposed by al-Tusi was via blue, which with increasing proportions of black becomes first sky blue and then turquoise, azure (*lazward*), indigo blue, and finally black-blue. In poems from al-Kindi's time, the relationship between blue and green was already being emphasized, as was the closeness of blue to dark brown. The fact that a distinction was seldom made between blue and green also explains how the legendary emerald-green mountain Kâf could give the sky its blue color.

AIR AND THE FORMS OF LIGHT AND COLOR

Two centuries after al-Kindi, a scholar named Ali al-Hasan Ibn al-Haytham lived in present-day Egypt. Al-Haytham, known in Europe as Alhazen since the Middle Ages, is considered the most important researcher in optics between antiquity and the seventeenth century. He made a significant contribution to our understanding of the visual process and systematically investigated the optical properties of air. This later fame notwithstanding, al-Haytham's biography, written by Ibn al-Qifti in the thirteenth century, shows how grueling the life of a scientist could be at that time. Al-Haytham was born around the year 965 in Basra (Iraq). Already renowned as a mathematician, he was invited by the caliph al-Hakim to come to Egypt. Al-Haytham had claimed he could build a dam that would control the Nile and thus guarantee a consistent harvest along its banks. But when he traveled up the Nile and saw the colossal construction projects completed in Pharaonic times, he realized that if such a thing were possible, a dam would already have been built back then. He withdrew his offer and fell out of favor with the caliph. Since al-Hakim was known to be unpredictable and murderous, al-Haytham feared for his life. In order to protect himself, he pretended to have gone mad, since according to the law madmen could not be executed. However, his possessions were confiscated and he was placed under house arrest. Not until al-Hakim's death in the year 1021 did al-Haytham, after twenty years' imprisonment, "miraculously" regain his sanity. He was rehabilitated and spent the rest of his life teaching and writing. He died in Cairo in the year 1040.

Whereas in al-Kindi's time the translation of Plato's and Aristotle's works had just begun, al-Haytham could already read all the important writings of the Greek philosophers in Arabic. This was certainly a great advantage, and it is beyond doubt that al-Haytham knew his Greek predecessors well. In his book *Optics* (Kitab al-Manazir), he refuted the extramission theory of vision, arguing so convincingly that hardly anyone was inclined to believe in the existence of visual rays ever again.

Al-Haytham saw himself as an experimental natural scientist who conducted experiments with light and its propagation. In his optical theory, he combined geometry and physics with the physiology of the visual process, taking the anatomy of the eye into account as well. This was an entirely novel approach to the study of vision.

The fundamental concept in his theory concerns the "forms" of light and color. By form (Arabic *sura*), he means something akin to what the Greek philosopher Democritus had designated as *eidola*: the images of objects that travel through a transparent medium to eventually stimulate perception in our eyes. He explains that all luminous and illuminated bodies radiate forms in every direction. For al-Haytham, as with Aristotle, it is a property of transparent media—such as air, water, and glass—that they conduct light and colors.

Al-Haytham does not consider air to be merely transparent, but also ascribes to it a certain slight density. He believes that this has a visible consequence:

> Therefore, when sunlight irradiates the air, it traverses the air in accordance with the air's transparency, and a small amount of the light is fixed in it in accordance with its slight density. Thus the light that is fixed in a small volume of air is very little because the volume of air is small and because air is very transparent and [only] a little dense and because the quality of the light that is fixed in it is weak.[2]

In other words, the air retains part of the light and is lit up by it. Conversely, direct sunlight grows weaker as it travels through the air. Both of these effects occur even when the air contains no haze particles. This explains the diffuse daylight that enables us to see our surroundings even on overcast days. Furthermore, the air causes objects at greater distances to appear relatively colorless:

> [W]hen the visible object moves away too far the form of its color fades and weakens. For it was shown that the form of color weakens as it recedes from the color from which it emanates, and that the same holds for the form of light.[3]

These words remind one of Aristotle. Al-Haytham also follows the Greek philosopher in writing that fumes can change the apparent colors of objects located behind them. Unfortunately, he does not say how. Al-Haytham concerns himself almost exclusively with the effects of air on light. He is much less interested in its effects on colors. His investigation of the rainbow, for example, deals with the shape in detail but largely ignores the sequence of its colors. The only thing he has to say about the sky's blueness is that it cannot be an optical illusion. The proof, he argues, lies in the fact that the color is reflected by water without changing appreciably.

While al-Haytham's theory offered little progress toward explaining why the sky is blue, one of his contemporaries made an insightful observation concerning the atmosphere. This man was Abu Rayhan Muhammad Ibn Achmad al-Biruni, a cosmopolitan astronomer employed by the caliph of Ghasni, in today's Afghanistan. From time to time he went on extensive journeys throughout the Middle East and into India, where he studied Sanskrit. On one of his first expeditions, probably in A.D. 995, the twenty-two-year-old went to Persia and climbed the country's highest mountain, Demavend (near today's Tehran). Two centuries later, the Persian scholar Zakarija al-Qazwini was to remark that Demavend "is so tall that it touches the stars, and resembles them in being unreachable."[4] Yet al-Biruni was undaunted and climbed it, a rare feat at the time. We do not know if he reached the summit. His account suggests that he reached a high enough altitude to notice that the clear sky was darker there than in the plains below (see Color Plate 5).[5] He assumes that this should be the case on every tall mountain. This observation points to the fact that skylight does not originate in a shining screen but rather is generated within the sphere of air, as Theophrastus and al-Kindi had correctly surmised.

THE BLACK AND THE PURE

While in al-Kindi and al-Haytham we have recognized scholars thinking about optics and the color of the sky, Ahmed Ibn Idris al-Qarafi was a thirteenth-century layman who addressed the same question. Al-Qarafi made his living by teaching Islamic law at a Quranic school in Cairo. An amateur scientist, he authored treatises on natural history, one of them on the "careful observation of that which the eyes see." In it he writes:

> Question: Why is the sky blue, when according to the as-
> tronomers it has no color?
> Answer: One does not see the sky at all, and what we don't see,
> we see as dark [without color]. Ask a blind man, for example,
> what is it that you see, and he will answer: "black darkness." And
> so the sky is also black, and under it is the air, which is transparent
> and luminous. Our gaze penetrates the air and sees it against the
> sky, so to speak. . . . Thus, the blue color (*lazward*) results from the
> purity of the air and the darkness of the sky, for we are dealing
> here with the mixture of the black and the pure.[6]

With a simple and direct application of Aristotelian color theory, al-Qarafi answers a question that had led al-Kindi to contemplate the state of the air and the nature of its particles. Presumably he had access to an Arabic edition of *On the Senses*, because this essay had already been translated in the ninth century. If that is the case, then al-Qarafi mistakenly conflates the mixing of black and white with the superimposition of light in front of dark in the sky, listed by Aristotle as two different ways to produce colors. After describing the overlapping of light in front of dark in the sky, he calls this a *mixture* of the Aristotelian base colors black and white ("the pure"). Al-Qarafi does not say how it is that air can be luminous, and thus circumvents the actual crux of the explanation, which had forced al-Kindi into an extended discourse.

When al-Qarafi died in Cairo in 1285, five hundred years had passed since the founding of the House of Wisdom. By now the golden age of Islamic science was in decline. In 1258 the Mongols had conquered Baghdad and destroyed the Abbasid caliphate. Christendom drove Islam completely out of Spain, which had been a center of Muslim culture for centuries. At the same time, skepticism toward science was on the rise throughout the entire Islamic world because of doubts that science could be reconciled with the teachings of the Quran. Many scientific treatises from this time contain apologetic disclaimers asserting that they are in harmony with the Quran and contribute toward the practice of the Islamic faith. The wars and religious rulers' distrust did little to create a climate for the carefree pursuit of natural philosophy.

If science was meeting with increasing resistance in the Islamic world, natural philosophy in Europe was just gaining momentum in the thirteenth century. When European scholars realized that their Arab colleagues had preserved the legacy of classical philosophy over the centuries, they eagerly awaited the translation of these works into Latin. Aristotle was about to return to Europe.

A SKY-BLUE SAPPHIRE

Before the first translations of Aristotle's works were circulated at the end of the twelfth century, European science had been in a deep crisis. There were almost no resources available for understanding the world philosophically. Most important was the Bible, augmented by a few Greek and Roman writings. Throughout the whole of medieval Europe, the only available classical text about the origin of colors was Plato's dialogue *Timaeus*, which had already been translated into Latin in the fourth century.

Instead of *explaining* the blue color of the sky, it was customary in the early and high Middle Ages to *interpret* this color as a symbol. The so-called allegoresis of precious gems had been a popular approach since the beginnings of Christianity. It associated the properties of the gems—that is, their color, luster, hardness, and form—with meanings, most of which were borrowed from the Bible. The sapphire, a blue and especially precious gem, was identified with the clear sky because of its color. An early reference can be found in the Book of Exodus in the Old Testament. When Moses seals the covenant with God on Mount Sinai, it reads as follows:

> Then went up Moses, and Aaron, Nadab, and Abihu, and seventy
> of the elders of Israel, and they saw the God of Israel, and under

his feet as it were a paved work of a sapphire stone, and as it were the body of heaven in His clearness.[7]

Here the sapphire is a symbol representing the throne of God. The symbol is used similarly by the prophet Ezekiel, who describes his vision of Heaven and the heavenly host as follows:

And above the firmament that was over their heads was the likeness of a throne, as the appearance of a sapphire stone, and upon the likeness of the throne was the likeness as the appearance of a man above upon it.[8]

The relationship of the sapphire to the sky was derived from these two passages of the Bible. The symbolism of the sapphire forged a link between the blue sky overhead and the religious hopes for heaven as the place of eternal life. If this gem was considered to be especially pure, the cloudless blue sky was likewise regarded as pure and immaculate. This is how the venerable Bede, an English monk of the eighth century, interpreted the sapphire. Later on, the meanings of this stone were expanded so that it stood at times for a life oriented toward heaven, for the right of Christians to become its citizens, and for the necessity of dissolving the bonds that tie human beings to an earthly life. The blue color of the Virgin Mary's robes in many medieval paintings also represents this meaning. The *Blue Virgin* of Chartres is a case in point (Color Plate 8). When this window was manufactured in a Chartres workshop in the late twelfth century, the intellectual climate in Europe was about to be transformed profoundly. Scholars began to discover the writings of Aristotle, and with them a theory that was far more consistent and versatile than that which Plato had laid out in *Timaeus*.

ARISTOTLE RETURNS

In contrast to the organized translating efforts of the House of Wisdom, the Latin editions of the twelfth and thirteenth centuries were the work of individual scholars, who often were completely unaware of each other.

CHAPTER 2

The most important translators of Aristotle's works were Gerard of Cre-
mona and William of Moerbeke. Over the course of about thirty-five
years, Gerard translated twelve astronomical and twenty-four medicinal
texts, seventeen books on mathematics and optics (including Euclid's *Elements* and al-Kindi's *Optics*), as well as fourteen works on natural philosophy (among them Aristotle's *Physics, On the Heavens,* and *Meteorology*).
Gerard specialized in translations from Arabic, while William used Greek
originals and tried to assemble a complete collection of Aristotle's works.
In the early thirteenth century, al-Haytham's *Optics* was rendered in Latin
by an unknown translator. Within a short time, copies of the new translations found their way into the major monasteries and the newly founded
universities.

The enthusiasm for Aristotle was so great that his works soon became
required reading at the universities, which supplanted the monasteries as
centers of cultural life in the thirteenth century. However, it did not take
long for his philosophy to come into conflict with church doctrine.
Aristotle was no Christian philosopher. To put it simply, he identified
God with the universe and considered His works to be absolute and immutable: how could there be miracles? Furthermore, he saw the universe
as an eternal formation without a beginning: what room did that leave
for a creator? In *Meteorology,* Aristotle had written that there could be no
water beyond the sphere of air, yet in the book of Genesis, right at the
beginning of the Bible, it says that God "divided the waters which were
under the firmament from the waters which were above the firmament."[9] In an attempt to reconcile the Bible with Aristotle, scholars modified the model of the world and inserted a sphere of water above the
fixed celestial sphere.

The church could hardly be mollified by such contrivances. Though
two archbishops, Albertus Magnus in Cologne and Robert Grosseteste
in Canterbury, were among the preeminent authorities on Aristotle in
the thirteenth century, even they could not prevent decrees from being
issued in the years 1210, 1215, 1231, and 1277 banning the teaching of
Aristotelian philosophy at the University of Paris. After the decree of
1231, the works of Aristotle did reappear in the curriculum as soon as
1240, and in spite of the ban against teaching his philosophy, his books

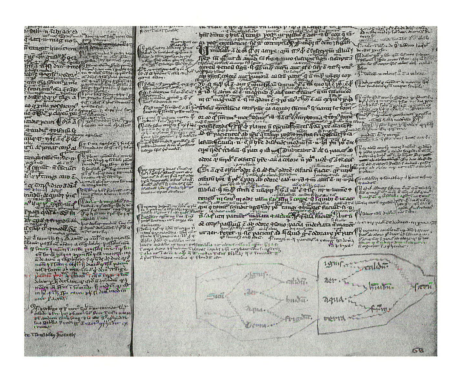

Figure 2.4 Aristotle's treatise *On the Soul* (De anima) served as an essential natural philosophical foundation for the classical and medieval understanding of colors. The many annotations in this thirteenth-century copy demonstrate how strong the interest in Aristotelian philosophy must have been. In the lower margin, a diagram depicts transformations of the four elements (see Figure 1.5). Courtesy of Staatsbibliothek Preussischer Kulturbesitz, Berlin; Ms. Lat. Qu. 341, fol. 66v–67r.

could still be read (Figure 2.4). And yet we can sense that, for scholars who wanted to continue the work of the Greek philosopher in the thirteenth century, it was not easy to steer clear of a distrustful church.

BLUE DEPTHS

Born in 1220, Roger Bacon fell victim to this distrust. The Englishman first taught at the Parisian university until the year 1247, when, at twenty-seven years of age, he moved to Oxford and there joined the Franciscan order. He began teaching at the university, but soon got into trouble with his superiors in the order. Perhaps his enthusiastic studies of Aristotle's works aroused suspicion. He was sent back to Paris in 1257 and placed under supervision. Although his superiors did recognize his talent and gave him the opportunity to carry on with his research, his findings led them to accuse him of heresy. He was forbidden to publish his ideas, and his manuscripts were sent to the papal court for inspection. Bacon appealed to Pope Clemens VI on the usefulness of the Aristotelian teachings and stated his goal of creating a comprehensive synthesis of natural philosophy. His main interest lay with optics, which he regarded, like al-Kindi before him, as the mother of all the sciences. Clemens was favorably impressed but died before he could approve Bacon's project. Often copied in secret, the monk's works were widely distributed throughout Europe during his lifetime, especially the *Great Work* (Opus maius), *On the Multiplication of Species* (De multiplicatione specieorum), and the *Third Work* (Opus tertium). Roger Bacon died in the year 1292.

Bacon was deeply impressed by al-Haytham, and his writings on optics owe more to the Arab scholar's work than to any other source except for Aristotle. What Democritus had referred to as *eidola* and al-Haytham designated as *forms*, Bacon calls *species*: the images of visible things. For him, as with Aristotle and al-Haytham, air is a transparent medium whose transparency determines its apparent color. He concurs with his two predecessors that the visibility of air is a consequence of its great spatial depth and its density. And he subscribes to the view of al-Haytham that an apparent darkening can be observed in transparent media with great depth. However, Bacon accounts for this darkening with two arguments that he cannot have inherited from his Arab mentor. Citing the example of water, which he designates as a transparent medium like air, Bacon says we can observe how the deeper layers are shaded by the shallower ones and therefore appear dark. He suspects that these shadows are cast by particles. The same goes for air across great depths.

Bacon cites the weakening of visual rays as a second cause of this darkening. This attempt to reconcile contradictory views reminds one of *Meteorology*, where Aristotle had surprisingly resorted to visual rays in order to explain the colors of the rainbow. Every dense body limits the visual rays (referred to here as *species*) as a function of its density, writes Bacon, adding that this must also occur in air, since its density is compounded over great distances:

> I say here that the air, or the sphere of fire, or the heavens, near and remote, is of similar rarity as far as perception is concerned; but it has, however, some density of its own nature, and this density is able in a great distance to terminate the species of vision, which it cannot do in a short distance, and therefore it will be quite visible at a distance, but not near at hand.[10]

According to Bacon, visual rays and light rays can reach that portion of a transparent medium which lies in close proximity to an observer unhindered, so that it appears brightened. At greater depths, however, the cumulative shading of the particles and the weakening of the visual rays result in increasing darkness. Only in a thick layer of air or in deep water does this darkness dominate the brightness of the close-up portion (Figure 2.5). For Bacon, Aristotle's color theory directly predicts what visible consequence this superimposition of lightness and darkness will have:

> Why a color appears approaching black, namely, blue (*aqua maris*), is explained in the same way as in the case of deep water, where in a like manner that color appears owing to shadows projected by particles. Darkness is caused by these shadows, which is similar to blackness. This is what takes place in the air or medium between us and the last heaven.[11]

Bacon clearly recognizes a common cause in the color of the sky and the color of water. His account is a thought experiment, since he could only speculate about the microscopic structure of these media. Like al-Kindi, he traces the blue color back to the effect of tiny particles in the air. But while the Arab scholar had assumed the presence of foreign bodies, the English monk argues that the blue color can be generated in pure air, provided that it has sufficient depth. With their notion of blue as "a color approaching black," Bacon, al-Kindi, and al-Qarafi are influenced by

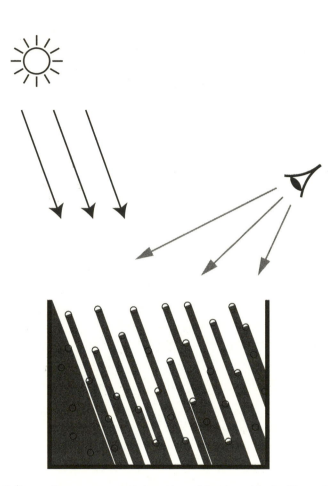

Figure 2.5 Roger Bacon compared the color of the sky with the blue color of deep water.

Aristotle's *On the Senses*. The origin of the color blue in the superimposition or mixture of light and dark appears to be the Aristotelian bottleneck for all early attempts at explaining the blue color of the sky.

THE PAINTER'S BLUES

In Tuscany one of Bacon's contemporaries found another way through the Aristotelian bottleneck: Ristoro d'Arezzo, also a monk. Probably born between 1210 and 1220 in Arezzo, Ristoro achieved fame only after his death, circa 1290. In a book from the year 1282 entitled *On the Composition of the World* (Della Composizione del mondo), Ristoro had tried to sum up all of the knowledge of his time with regard to the makeup of the world. As the first book on natural science written in the Italian language, it remained popular in Italy throughout the late Middle Ages.

The Composition of the World is a compendium of older views on cosmology. Ristoro refers primarily to Aristotle *(Meteorology* and *On the Heavens)* and his Arab commentator Averroes. Without a doubt, he was familiar with Ptolemy's *Optics*, as well as with a series of Arab authors. But he also incorporated his own thoughts into the book. This is clearly evident in its chapter on the blue color of the sky. Although Bacon and Ristoro had essentially the same sources at their disposal, their explanations of the sky's color differ considerably. Ristoro explains it through an analogy between nature and the painter's practice:

> According to the statements of scholars the sky is actually supposed to be colorless; let us then contemplate the reason why it appears blue *(azzurro)*. Clever painters who paint in color—when they want to get the color blue—mix two different colors together: light and dark, and from this mixture, blue results. When I look at the sky, I see two opposite colors mixed, namely light and dark, and this is due to the air's depth.[12]

Even Aristotle, in *On the Senses*, mentions the relationship between the colors of the atmosphere and painters' color-mixing techniques. And we know that since ancient times, one of the tricks of the painter's trade had been the combination of black and white pigments to produce a gray-blue color. To achieve this, they applied translucent black pigments to a white base coat. This technique was widely used up until the Renaissance, and it can even be found in the work of the Venetian painter Tizian in the sixteenth century. However, a blue color results only when the pigments are *superimposed*, and not when they are mixed, as Ristoro claims.

In contrast to al-Kindi and Roger Bacon, Ristoro does not go into detail on how air can be brightened. Rather, he concentrates on identifying what is dark and light in the sky, and insists that they correspond

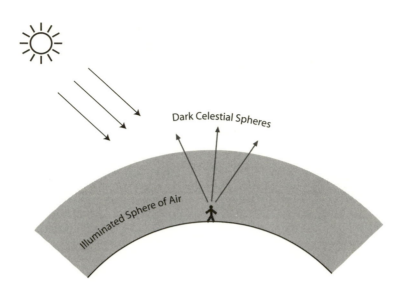

Figure 2.6 Ristoro d'Arezzo thought the blue color of the sky was caused by the dark heavenly spheres shining through the illuminated sphere of air.

with the light and dark colors in paintings. In his Aristotelian cosmology, the dark is easy to identify: it is in the dim sphere of fire and the heavenly sphere. If we look at the clear sky, our gaze necessarily ends in these dark spheres, but not until it has traversed the atmosphere (Figure 2.6). During the day the sun illuminates the atmosphere, and thus brightens it. It seems natural to Ristoro to identify this scenario with the light colors of painters:

> When I look at the sky and gaze with my eyes on this darkness that lies beyond the light, I see how the light mixes with the dark along the way, like muddy water when it rises up into the clear water above. And to the eye, the result of this mixture of dark and light is the color blue (*colore d'azzurro*).[13]

In concluding his chapter, Ristoro addresses the claim that the color of the sky results from blue-colored air. This must have been a popular explanation at the time, since otherwise Ristoro would not have taken the trouble of arguing against it. He offers a convincing reason to doubt this

notion. Ristoro compares the effect that could be expected from colored air with the familiar effect of colored glass. If we look through the
latter, all of the objects behind it appear to have the color of the glass, or
at least a mixture of their own color with that of the glass. If the sky were
blue because the air is blue, then everything behind the atmosphere
would have to appear blue in color. The sun, the moon, and all the stars
would look blue! Ristoro knew that this is not the case: the sun looks
white or yellowish during the day and reddish as it sets. Moreover, it had
been pointed out since antiquity that many stars are colored. Some of
the fixed stars appear blue, while others are white or reddish. And even
the colors of some wandering stars—the planets—were known in the
thirteenth century: Venus is pale green, Mars is red, and Saturn is yellowish. Properly interpreted, such simple observations yield a profound insight into the nature of the air.[14]

Aerial Perspective

F riday, November 11, 1994. The tension is rising at Christie's, New York. This company's auction room on Park Avenue is packed with people, buzzing with excitement. In the morning, a catalog of rare books has been auctioned, but now bids are being accepted for the day's high-light: catalog number 8030, a manuscript of eighteen weathered leaves, with seventy-two pages altogether. Each page is filled with densely scrib-bled handwriting, in mirror style from right to left. Here and there ink drawings dot the pages, depicting bridges, the flow of water in rivers, and the illumination of the moon. The bid started at $5.5 million. Meanwhile, it has passed $10 million and keeps rising steadily.

Catalog number 8030 is Leonardo da Vinci's *Codex Leicester*, and de-spite their inconspicuous appearance, its pages attest to the scope and vi-sion of a Renaissance genius. The *Codex* contains Leonardo's notes from a journey into the Alps, undertaken from Milan sometime between 1508 and 1510. On this journey, Leonardo (Figure 3.1) studied the cir-culation of water in the cosmos, from its evaporation on Earth, the for-mation of clouds and downpours, to the ways rivers enter the sea. The notebook also deals with the moon's so-called ashen light. This long-standing riddle of astronomy concerned itself with the question of why the disk of Earth's companion is faintly visible a few days before and af-ter new moon, when a thin, bright sickle is all of the lunar surface that is illuminated by the sun directly. Leonardo's answer, first put down in the *Codex*, remains valid today: it is sunlight reflected off Earth, which illuminates the dark side of the moon and makes it visible to our eyes. Among the other subjects contemplated by Leonardo in the *Codex* is the color of the daytime sky.

Named after Thomas Coke, the first Lord Leicester, who had bought it from an Italian painter in 1717, the *Codex* had remained in the lord's

Figure 3.1 Self-portrait of Leonardo da Vinci in red chalk from around 1516.
Biblioteca Reale, Torino. Courtesy of Bildarchiv Preussischer Kulturbesitz, Berlin.

library at Holkham Hall in England for 263 years. But business was going badly for Coke's heirs, and so, in 1980, they put up the *Codex* for
auction at Christie's in London. It fetched $5.6 million then. The buyer
was Armand Hammer, a flamboyant Italo-American oil magnate and
collector of art. As the story goes, Hammer had the *Codex* guarded by a
squad of kung fu fighters. He died in 1990, and soon afterward a law-
suit broke out over the ownership of his collection. Now, in 1994, the
Codex is being sold to pay the legal expenses of the Armand Hammer
Art Museum in Los Angeles.

Before long, the bid has passed the $20 million mark. Only two bid-
ders remain. One of them is Cariplo, a bank from Milan. Representa-
tives of this company have declared the return of the *Codex* to Italy an
issue of national importance. The other is a private collector who is not
present in person but is participating in the auction via telephone. In the
end, Cariplo's efforts are in vain. At $30.8 million, the gavel goes down,
and the *Codex* is sold to the private bidder. The next day, this person is
revealed to be William Gates III, the chairman of Microsoft Corpora-
tion. The *Codex* remains the only Leonardo manuscript still in private
hands and the most expensive manuscript sold to date.[1]

When he composed the *Codex Leicester*, almost five hundred years
prior to the New York auction, Leonardo was in his late fifties. He had
already painted all of his famous paintings, such as the *Mona Lisa* and
The Last Supper, yet his curiosity about nature was as fresh as ever. The
son of a maid and a successful notary, Leonardo was born in 1452 in
Anchiano, a small village near the town of Vinci in Tuscany. It is said
that, as a little boy, Leonardo went on expeditions around Anchiano in
order to observe lizards, glow-worms, and beetles. His love of nature
became evident when he bought live birds at the market in Vinci only
to let them fly away, restoring their freedom. Leonardo's artistic inclina-
tion also seems to have emerged early in his youth. When his father
moved to Florence in 1469, he took his son to Andrea del Verrocchio to
be educated in this man's famous workshop. Verrocchio was not only a
masterful painter, sculptor, and goldsmith, he was also well-versed in
mathematics, geometry, and optics. While Leonardo was thus introduced
to the optical tradition, his experiences as an apprentice in an artist's

studio exerted a far greater influence on his work at first. After remaining with Verrocchio for four years, he opened his own workshop as an accredited Florentine painter. In 1482 Leonardo left Florence, following an invitation from Ludovico Sforza, the Duke of Milan, to develop military machines and to erect a huge monument for his employer. Like many of his later works, it was never finished. While in Milan, Leonardo began to record his thoughts and observations in his notebooks.

Throughout his life, Leonardo kept his distance from academic scholars. Often he ridiculed them as blatantly rehashing the works of others while praising himself for drawing his wisdom directly from personal experience:

> If indeed I have no power to quote from authors as they have, it is a far bigger and more worthy thing to read by the light of experience, which is the instructress of their masters. They strut about puffed up and pompous, decked out and adorned not with their own labours but by those of others, and they will not even allow me my own. And if they despise me who am an inventor how much more should blame be given to themselves, who are not inventors but trumpeters and reciters of the works of others?[2]

Leonardo is understating his own merits here, since over the years he gained an impressive knowledge of the scholarly tradition. His notebooks reveal his readings of Aristotle's *Meteorology*, Theophrastus' *On Colors*, the *Major Work* of Roger Bacon, John Pecham's *Perspectiva communis*, and Ristoro d'Arezzo's *On the Composition of the World*. These are precisely the writings that had set the course for optical inquiries throughout the Middle Ages. Leonardo tried hard to make his own observations agree with the optical tradition defined by these texts.

THE BLUE ELEMENT

What was the initial source of Leonardo's understanding of color? Leon Battista Alberti's book *On Painting* (Della pittura) is the best candidate. It

dates to around 1435. Alberti had studied law and liberal arts in Bologna
before returning to his hometown of Florence, where he soon became
a renowned architect, sculptor, painter, and art theorist. At this time
the Medici family generously supported the arts, sciences, and architec-
ture. Alberti succeeded in becoming a wealthy and influential architect.
His *On Painting* was originally meant to be a manual for artistic practice. In retrospect, it is considered the starting point for Renaissance art
theory. Aristotle's philosophy had dominated the university curriculum
in Bologna, and its influence is clearly evident in *On Painting*. Nevertheless, when dealing with the systematics of color, Alberti does not follow the linear scale that the Greek philosopher had proposed, considering it to be impractical for painting. As an alternative, Alberti suggests
that the four elements earth, water, air, and fire are tinged in characteristic colors:

> Fire is of the color called red. Then there is air, which is said to be
> sky color (Latin: *caelestis*, Italian: *celestrino*) or grey-blue. The color
> of water is green, and earth has the colors of ashes. We see all the
> rest of the colors made from a mixture of these. . . .[3]

In the fifteenth century, such an association between colors and the
four elements was far from novel. Rather, it alluded to a tradition that
can be traced back to Empedocles in Greek antiquity. Since then, numerous thinkers had tried to associate colors with the elements, a cus-
tom continuing through the Renaissance (Table 3.1). Alberti seems to
have been the first to relate the sky's blue color to the element air. Previously, Theophrastus and Isidore of Seville had declared air to be white,
and Empedocles had even identified it with a yellowish green (*ochron*)
or red.

All matter is composed of the four elements, Alberti reasons, and
therefore the color of any object will reflect the proportions of its constituent elements. To paint the color of an object as a mixture of the four
primary colors thus requires an understanding of its elemental composition. Art should seek to imitate nature, he writes, but it can do so only
when the painter understands the material constitution of natural bodies.
This relationship between artistic practice and the observation of nature,

TABLE 3.1

ATTEMPTS AT ASSIGNING NATURAL COLORS TO THE FOUR ELEMENTS EARTH,
WATER, AIR, AND FIRE HAVE HAD MIXED RESULTS

Earth	Water	Air	Fire	Investigator
Yellowish green (*ochron*) or red	Black	Yellowish green (*ochron*) or red	White	Empedocles (c. 450 B.C.)
White	White	White	Yellow	Theophrastus (c. 300 B.C.)
Black	White	Red	Yellow	Galen (c. 180 A.D.)
The colors of ashes	Green	Blue	Red	Leon Battista Alberti (1435)
Yellow	Green	Blue	Red	Leonardo da Vinci (c. 1492)
Blue-green	Violet	Yellow	Red	Mario Equicola (1526)
White	Green	Blue	Flame color (*flavo*)	Joseph Justus Scaliger (c. 1600)
Black	Blue	Yellow	White	Athanasius Kircher (1646)

SOURCES: Empedocles: Hermann Diels, *Doxographi Graeci* (Berlin: G. Reimer, 1879), 315.

Theophrastus (Pseudo-Aristotle): *De coloribus* 793b34–794a15; Aristotle, *Minor Works*, trans. W. S. Hett, Loeb Classical Library (Cambridge, Mass.: Harvard University Press, 1952), 5.

Galen: John Gage, *Color and Culture: Practice and Meaning from Antiquity to Abstraction* (Boston: Little, Brown and Co., 1993), 29.

Alberti: Samuel Y. Edgerton Jr., "Alberti's Color Theory: A Medieval Bottle Without Renaissance Wine," *Journal of the Warburg and Courtauld Institutes* 32 (1969): 122.

Leonardo: Maria Rzepinska, "Leonardo's Colour Theory," *Achademia Leonardi Vinci* 6 (1993): 19.

Equicola and Scaliger: Leon Battista Alberti, *Kleinere kunsttheoretische Schriften*, trans. into German by Hubert Janitschek (Vienna: Wilhelm Braumüller, 1877), 64.

Kircher: Martin Kemp, *The Science of Art: Optical Themes in Western Art from Brunelleschi to Seurat* (New Haven: Yale University Press, 1990), 280.

as described by Alberti, came to be a pillar of Leonardo's thought some forty years later.

Besides color systematics, the increase in darkness as a function of distance had received much attention in optical theories during antiquity and the Middle Ages. We saw in the last chapter how this was alternatively explained by the weakening of light or the existence of visual rays. Alberti considers darkening to be of great importance in painting. He advises artists to paint objects that are meant to seem near in a mixture of their true colors and white, but distant ones with an increasing admixture of black. This, Alberti holds, will create the impression of spatial depth in a painting. For example, a green tree should look dark green in the distance and light green when seen from close up. While *On Painting* explicitly promoted this as a standard technique for painting, the idea was not entirely new. Some artists had applied it many years before. A noteworthy example can be found in the frescoes of Giotto di Bondone in the Scrovegni Chapel in Padua, painted between 1305 and 1313.

By introducing this rule, Alberti set himself apart from Aristotle. The Greek philosopher assumed distant objects to appear in the dark colors green, violet, and blue, whereas closer ones appear in the brighter colors yellow and red. While for Aristotle distance affected an object's apparent hue, for Alberti only its brightness was affected. Unlike Aristotle, Alberti thus distinguished between the hue and the brightness of color, a necessary condition for the so-called aerial perspective that Leonardo was to develop half a century later.

BLUE AT A DISTANCE

In claiming that air is tinged blue, Alberti ostensibly overcame all the hurdles in explaining the sky's color. Throughout the Renaissance, Aristotle's model of a spherical world was widely accepted, and the assumption of spheres of water, air, and fire extending above Earth's surface was considered solid knowledge. Therefore, the color seen in the sky would appear to be a logical consequence of the blue element filling the sphere

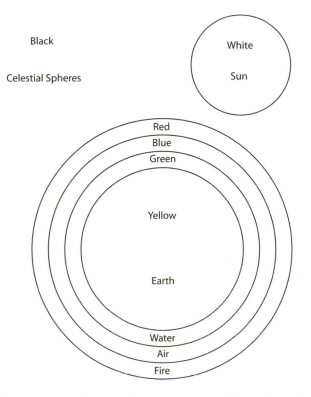

Figure 3.2 Following Leon Battista Alberti, Leonardo da Vinci related the colors to the four elements in 1492. After Martin Kemp, *The Science of Art* (New Haven: Yale University Press, 1990), 268.

of air. Alberti must have been unaware that this thesis had been refuted much earlier by Ristoro d'Arezzo.

Leonardo, in one of his first remarks on color (c. 1492), relies on Alberti's association of the elements with the primary colors and assumes blue to be the natural color of air (Figure 3.2). He writes:

White is given by light, without which no color may be seen, yellow by earth, green by water, blue by air and red by fire, and black by darkness which stands above the element of fire, because there is no substance or dimension on which the rays are able to percuss and accordingly to illuminate it.[4]

Figure 3.3 Sketch by Leonardo illustrating the
rule of aerial perspective, drawn between 1487
and 1490. Reproduced from Jean Paul Richter,
The Literary Works of Leonardo da Vinci (London:
Phaidon Press, 1883), 1:159.

Except for earth, which Alberti deemed ash-colored, these are exactly the
relations between the colors and the four elements stated in *On Painting*.
Starting from this premise, Leonardo wonders how blue air would affect
the colors seen in nature:

> The medium which lies between the eye and the thing seen
> transmutes that thing into its own color. For example, the blue air
> makes distant mountains appear blue; a red glass makes whatever
> the eyes see through it look red.[5]

The air thus acts like a blue color filter, making objects appear increasingly blue-tinged with growing distance. Because of his interest in the
realistic pictorial depiction of landscapes, this finding is of capital importance for Leonardo, who translates it into a rule for painting. In depicting a landscape an artist should progressively increase the admixture of
blue in the colors of its distant parts. He illustrates this rule with advice
on how to paint buildings at different distances (Figure 3.3):

> [T]hrough the variety of air one can recognize the different dis-
> tances of various buildings which appear to be severed from their
> bases along a single line. This would be the case if many buildings
> were seen beyond a wall, so that all of them appeared to be of the
> same size above the edge of that wall. And if you wanted, in paint-
> ing, to make one seem more distant than the other, it would be
> necessary to represent the air as a little thick, for you know that in
> air like this, the farthest things seen in it, such as mountains, seem
> blue—almost the color of the air when the sun is in the east—
> because of the great quantity of air that is found between your eye
> and the mountain. Therefore paint the first building above the

wall in its true color; the next in distance make less distinctly out-
lined and bluer; that which you wish to seem more distant, paint
as even bluer; and whatever you wish to show as five times more
distant, make five times bluer.[6]

Did Leonardo infer this rule from actual observations of landscapes?
Probably not. Considering the way Alberti's *On Painting* influenced his
thinking, it seems more likely that Leonardo derived it from the theoretical assumption of blue-tinged air. In fact, the appearance of distant landscapes is more complicated, and Leonardo was well aware of this. Even
though mountains in the distance do seem to be tinged blue owing to
the air interposed between them and an observer, they also appear lighter
in color, particularly at their base (Color Plate 9):

> Hence, O Painter! when you represent mountains, see that from
> hill to hill the bases are paler than the summits, and in proportion
> as they recede beyond each other make the bases paler than the
> summits; while the higher they are the more you must show of
> their true form and color.[7]

The air thus affects the appearance of distant mountains in three ways:
they appear bluish with increasing distance, while at their base they
seem more pale and blurred than toward their summits. These claims,
formulated by Leonardo as a set of rules, became known as the aerial or
color perspective (the bluish tint and the paling near the horizon) and
the acuity perspective (the blurring). Together with linear perspective,
which uses one or more vanishing points to create a three-dimensional
impression by representing distant objects as progressively smaller, they
became typical features of Renaissance painting. The most natural effect could be created when all three types of perspective were applied
simultaneously.

These studies had a noticeable influence on Leonardo's own artistic
practice. One famous example is his painting *The Virgin of the Rocks*
(c. 1506), a work done thirty years after his first studies of the optical
properties of air (Color Plate 10). The painting shows both the brightening and the blueing in the distance.

Leonardo was not the only member of Verrocchio's workshop to use these effects to suggest increasing distance in his paintings. His teacher Verrocchio himself had done so in a painting from c. 1467, even before Leonardo became his pupil, and other Florentine painters of the 1460s and 1470s did so as well. Interestingly, this practice deviated from Alberti's *On Painting*, which had only recently become accepted as the authoritative work on painting. Perhaps this deviation reflects the innovative atmosphere that prevailed in Florence's art community during this period. Nevertheless, the background landscape in *The Virgin* shows how Leonardo transcended the pictorial conventions of his time. In the faraway landscape behind the rock cave we can see two chains of mountains and a little patch of sky. While the rocks just behind the cave show their corporeal colors, the mountains seem to disappear in the distance. The first mountain of the chain has a bluish tinge, but still reveals its surface features. In contrast, the continuation of the chain in the distance is skillfully emphasized by progressive brightening and blurring. The transition toward these distant mountains is not abrupt but gradual. Thus, in painting this scene Leonardo simultaneously applied the aerial perspective and the acuity perspective. Nowhere do we see objects "proportionately bluer" with increasing distance, as he had advised before. However, since the distant mountains are situated closer to the horizon, the brightening may indeed surpass this bluish tinge. The brightening of the cloudless, blue sky near the horizon is discernible as well. In painting it, Leonardo tried to use his rules of perspective without compromising his artistic freedom. Looking at some of his other paintings from this period—such as the famous *La Gioconda* (*Mona Lisa*)—we discover the increasing complexity of his backgrounds. These often consist of rugged landscapes, which gave him plenty of opportunity to experiment with the effects of light, shadows, and colors.

IN THE FOOTPRINTS OF ARISTOTLE

Leonardo could only develop a systematic theory of aerial perspective by carefully studying the structure and composition of the sphere of air.

Aristotle's *Meteorology* continued to be the authoritative text on this topic. In 1490, Leonardo owned a copy of this book, according to a catalog of his library that he compiled at that time. By reading *Meteorology*, Leonardo learned that the density of air decreases with increasing elevation. His notes prove that he considered dense air illuminated by the sun to appear brighter than thin air, and that air makes distant objects appear blurred. Both views are reminiscent of al-Haytham's *Optics*, which Leonardo may have heard of through John Pecham's *Perspectiva communis*, a popular optics textbook of the late Middle Ages that he kept in his library beside Aristotle's *Meteorology*. For Leonardo, the increasing density of air near the ground, combined with its illumination by the sun, explained the brightening of distant landscapes near the horizon.

His knowledge of Aristotle and al-Haytham gave Leonardo second thoughts about Alberti's blue air, since both the Greek philosopher and the Arab scientist considered air to be colorless. Leonardo must have been torn between Alberti's claim, which seemed useful for painting, and the customary natural philosophy of his day. It appears that around 1490 his new familiarity with the optical tradition led him out of this dilemma and gradually made him reject Alberti's color doctrine. In a manuscript from this period we read:

> A dark object will appear more blue when it has a larger amount
> of luminous air interposed between it and the eye, as may be seen
> in the color of the sky.[8]

In mentioning the "luminous air" and the role of background objects, Leonardo alludes to al-Haytham, who had dealt at length with the effects of air illumination, while only vaguely linking it to the sky's color. Three years later, in 1493 or 1494, Leonardo writes:

> The air is blue because of the darkness which is above it, for black
> and white together make blue.[9]

This condensed Aristotelian explanation of the blue sky reminds one of Ristoro d'Arezzo's *On the Composition of the World*. Throughout the fourteenth and fifteenth centuries, this was the most popular book on cosmology in Italy. Like Ristoro, Leonardo emphasizes the necessity of a dark background to make the blue color of the sky visible (Ristoro had probably been alluding to Theophrastus' *On Colors* in doing so). Moreover, Leonardo's growing doubt over Alberti's blue air is evident, since here he considers the air to be white, just as Theophrastus had done. Having left the framework of Alberti's doctrine he now adopts, or rather returns to, the Aristotelian concept of color in claiming that blue results from a mixture of black and white. Leonardo experts believe that he never read Aristotle's *On the Soul* or *On the Senses*, and thus he must have learned about the Greek philosopher's ideas on color from the Aristotelian tradition, rather than directly from the writings of its master.

Alberti and Ristoro, two of Leonardo's early sources, refer to the practice of painting in relation to the coloration of natural objects. Yet, while Alberti presents a theoretical viewpoint, Ristoro claims his work relates directly to the practice of painting. The latter approach must have been to Leonardo's taste, since it affirmed his ambition of understanding the works of nature through painting. Just a couple of years before, he had carefully studied Alberti's doctrine as well. It thus appears that, in the 1490s, Leonardo still lacked a framework for the systematics of colors that would satisfy his own demands.

A JOURNEY TO THE ALPS

Leonardo had done most of these early studies on aerial perspective in Milan. When the Sforza dynasty was expelled from the city in 1499, Leonardo returned to Florence and was employed at the court of the Medici. It was during this Florentine period that he painted the portrait known as *La Gioconda* (*Mona Lisa*) and produced an immensely large mural in the Palazzo Vecchio, the town hall of Florence. Called *The Battle of Anghiari*, the painting celebrated a Florentine victory over Milanese troops in 1440. It was the largest public work Leonardo was ever commissioned with. Even as it was being painted, contemporaries regarded it as a major work and made pilgrimages to see it. Several artists copied it, and it is only through these paintings and Leonardo's preparatory sketches that

Figure 3.4 South side of Monte Rosa, on the border between Switzerland and Italy. Courtesy of Stefan Vehoff.

we know what it looked like, for it was subsequently painted over. Shortly after finishing the painting in 1506, Leonardo returned to the court of the Duke of Milan, the heir of those he had just depicted in defeat.

More than a decade after his first studies, Leonardo was still troubled by the question of whether air is blue (Alberti), colorless (Aristoteles and al-Haytham) or—on illumination—white (Ristoro d'Arezzo). But now there was new hope for overcoming this problem. In the interim, Leonardo had gained access to a Latin copy of *On Colors*, in which Theophrastus speculates on the sky's blue color and claims that air becomes white on illumination, just as Ristoro would much later. What is more, Leonardo had the opportunity to travel from Milan to the Alps one summer, probably in the year 1508, 1509, or 1510. He went to the Monte Rosa, a range of mountains on the border between Italy and Switzerland (Figure 3.4). During his travels, Leonardo carried out a series of thorough

investigations into the circulation of terrestrial waters, the light reflected
from the lunar surface, and the color of the sky. It is in the *Codex Leicester*,
the notebook of this journey, that Leonardo's previous studies meld with
his new observations to form a synthesis full of remarkable insight. For
purposes of orientation within these long texts and for reference during
the discussion that ensues, I have numbered the paragraphs. On a page
entitled "On the color of the air" (Del colore dell'aria), Leonardo writes:

1 I say that the blue which is seen in the air is not its own color,
but is caused by the heated moisture having evaporated into the
most minute imperceptible particles, which the beams of the solar
rays attract and cause to seem luminous against the deep intense
darkness of the region of fire that forms a covering above them.
And this may be seen, as I myself saw it, by anyone who ascends
Mon Boso [Monte Rosa], a peak of the chain of Alps that divides
France from Italy, at whose base spring the four rivers which flow
as many different ways and water all Europe, and there is no other
mountain that has its base at so great an elevation. . . . And I saw
the air dark overhead, and the rays of the sun striking the mountain had far more brightness than in the plains below, because less
thickness of air lay between the summit of this mountain and the
sun.
2 As a further example of the color of the air, we may take the
case of the smoke produced by old dry wood, for as it comes out
of the chimneys it seems to be a pronounced blue when seen between the eye and a dark space, but as it rises higher and comes
between the eye and the luminous air, it turns immediately to an
ashen grey hue, and this comes to pass because it no longer has
darkness beyond it, but in place of this, the luminous air. But if
this smoke comes from new green wood, then it will not assume a
blue color, because, as it is not transparent, and is heavily charged
with moisture, it will have the effect of a dense cloud which takes
definite lights and shadows as though it were a solid body.
3 The same is true of the air, which excessive moisture renders
white, while little moisture acted upon by heat causes it to be

dark and of a dark blue color; and this is sufficient as regards the definition of the color of the air, although one may also say that if the air had this transparent blue as its natural color, it would follow that wherever a greater quantity of air came between the eye and the fiery element, it would appear of a deeper shade of blue, as is seen with blue glass and with sapphires, which appear darker in proportion as they are thicker. The air, under these conditions, acts in exactly the opposite way, since where a greater quantity of it comes between the eye and the sphere of fire, there it is seen much whiter, and this happens towards the horizon; and in proportion as a lesser amount of air comes between the eye and the sphere of fire, so much the deeper blue does it appear, even when we are in the low plains. It follows therefore, from what I say, that the atmosphere acquires its blueness from the particles of moisture which catch the luminous rays of the sun.

4 We may also observe the difference between the atoms of dust and those of smoke seen in the sun's rays as they pass through the chinks of the walls in dark rooms, that the one seems the color of ashes, and the other—the thin smoke—seems of a most beautiful blue. We may also see in the dark shadows of mountains far from the eye that the air which is between the eye and these shadows will appear very blue, and in the portion of the mountains which is in light, it will not vary much from its first color.

5 But whoever would need a final proof should stain a board with various different colors, among which he should include a very strong black, and then over them all he should lay a thin transparent white (*biacca*), and he will then perceive that the lustre of the white will nowhere display a more beautiful blue than over the black,— but it must be very thin and finely ground.[10]

A few pages later Leonardo convincingly emphasizes the necessity of a dark background to make the blue color visible:

6 Experience it is that shows how the air has darkness behind it and yet appears blue. Make smoke of dry wood in a small quantity; let the rays of the sun fall upon this smoke, and behind it

place a piece of black velvet, so that it shall be in shadow. You will then see that all the smoke which comes between the eye and the darkness of the velvet will show itself of a very beautiful blue color; and if instead of the velvet, you put a white cloth, the smoke will become the color of ashes. An excess of smoke acts as a veil, a small quantity of it does not render the perfection of this blue: it is by a moderate admixture of smoke therefore that the beautiful blue is created.

Water blown in the form of spray into a dark place, through which the solar rays pass, produces this blue ray; and especially when this water has been distilled; and the thin smoke becomes blue. This is said in order to show how the blue color of the air is caused by the darkness that is above it; and the above-mentioned instances are offered for the benefit of anyone who cannot confirm my experience on Mon Boso.[11]

The systematic style of these notes suggests they were part of a book manuscript Leonardo was preparing. Ever since composing his early Milanese manuscripts, he had announced he was preparing a book on light, shadow, and color. Yet at the same time, his style is convoluted. Some of his favorite subjects do appear: the bluish tinge in the distance, the study of smoke, and painting as an instrument for perceiving the workings of nature. But he refrains from distinguishing his literary sources from what he observed himself.

While Leonardo claims to have ascended Monboso (Monte Rosa), it cannot be assumed that he reached its summit. Surrounded by glaciers, Monte Rosa is the second-highest peak in the Alps, with a height of 4,630 meters. Nevertheless, he may have reached a considerable elevation on its slopes, for he notes that at a high elevation, the cloudless, blue sky seems darker than on the plains below.

In contrast to his earlier notes of 1493–94, Leonardo is no longer content with simply reiterating Ristoro's blue as a mixture of light and darkness. Rather, he now locates light and darkness in the spheres of air and fire, respectively. He assumes that the illumination of imperceptible, moist particles by solar rays makes air appear white. When looked at in front of the dark sphere of fire, this mixture of light and darkness should give rise to the color blue, Leonardo claims. This is strongly reminiscent of Theophrastus, who reasoned in *On Colors* that dense air is white, but looks blue if penetrated by profound darkness. Yet Theophrastus had neither claimed that the air consists of "imperceptible particles" nor explicitly identified the sphere of fire as the region of darkness.

In the first paragraph, Leonardo refutes Alberti's claim that air is tinged blue. Instead, he builds on an Aristotelian foundation. In *Meteorology*, Aristotle had written of the sphere of fire's darkness and of the warm, moist vapors floating in the air and making it look white when illuminated by the sun. The Greek philosopher had considered air to be a combination of the qualities hot and moist, while arguing that water was a combination of the qualities cold and moist. The evaporation of heated moisture into imperceptible particles (1) thus marks a transitional state between air and water. The condensation of air seemed necessary to Aristotle in order to make colors visible in the sphere of air. These condensed vapors, the Greek philosopher wrote in *Meteorology*, act like small, airborne mirrors that are capable of reflecting colors, thus giving rise to the rainbow. It seems that Leonardo drew inspiration from Aristotle's explanation of the rainbow in accounting for the blue color of the sky (1 and 3).

Having concluded that the sky's blue originates in evaporated, moist particles that are illuminated by the sun, Leonardo perceives the same color in other media as well. Examples are smoke from old, dry wood (2) and "water blown in the form of spray" (6). Both Roger Bacon and Ristoro d'Arezzo had used vivid comparisons to explain the origin of the blue color. Leonardo goes far beyond these. His examples not only serve as intuitive illustrations, they also link the abstract color of the sky to everyday experience. His experiments can be repeated by anyone. The coloration of illuminated smoke provides Leonardo with proof that a dark background is needed to make the blue color appear (Table 3.2 and Color Plate 11). Extending this finding to the situation in the atmosphere, Leonardo concludes that the dark sphere of fire is a necessary ingredient in the formation of the sky's blue color.

TABLE 3.2

LEONARDO'S OBSERVATIONS OF THE COLORS SEEN WHEN DIFFERENT
SUBSTANCES ARE ILLUMINATED BY SUNLIGHT, AS DESCRIBED
IN THE *CODEX LEICESTER* (C. 1508)

Substance Illuminated by Sunlight	Background	Apparent Color
Smoke produced by old, dry wood	A dark space	A pronounced blue
Smoke produced by old, dry wood	Luminous air	Ashen-gray hue
Smoke produced by fresh, green wood		Effect of a dense cloud
Atoms of dust	Dark rooms	The color of ashes
Thin smoke	Dark rooms	A most beautiful blue
Air	Dark shadows of mountains	Very blue
Air	Portion of the mountains that is in light	Colorless
Small quantity of smoke from dry wood	A piece of black velvet in shadow	A very beautiful blue color
Small quantity of smoke from dry wood	A white cloth	The color of ashes
Water blown in the form of spray	A dark place	Blue

SOURCE: *Codex Leicester*, fol. 4r and fol. 36r; Edward MacCurdy, *The Notebooks of Leonardo da Vinci* (London: Jonathan Cape, 1938), 1:418–21.

Reading Leonardo's comprehensive expositions, one could almost forget that they were written by a master painter. There is only one paragraph (5) in which he connects the sky's blue color with artistic practice. In claiming that the superposition of a transparent white onto a black background yields a blue, he seems to allude to his own painting technique. However, it turns out that this must have been a mere thought experiment by Leonardo. Painters could indeed superimpose a fine layer of black pigment onto a white background to yield blue, but the reverse is much more difficult. And with the lead white (*biacca*) mentioned by Leonardo, this effect is impossible, since it is an especially dense and concealing pigment. As such, it is particularly unsuited for this technique. Attempts by art historians to reproduce the effect described by Leonardo have failed to yield blue, always resulting in a grayish green. A careful examination of *The Virgin of the Rocks* (Color Plate 10) confirms the suspicion that he did not follow his own suggestions. In painting the sky on the left-hand side, Leonardo applied a progressively tempered mixture of white and azurite (a blue mineral pigment) onto a white background. Thus, he does not produce the described effect in painting, but imitates it with pictorial means.

THE INFINITE DARKNESS

Leonardo understood that the daytime sky can only appear blue because of the darkness extending beyond the sphere of air. From a layman's point of view, the presence of this darkness may seem trivial. In daytime, the sun is up and shining. At night it has set, so obviously, it must be dark! Like most of the thinkers operating within the framework of Aristotelian cosmology, Leonardo does not consider darkness at night a problem. In his worldview space is finite, and even though there are some planets and many stars in the universe, the amount of light they radiate is small when compared with the intrinsic darkness of the celestial spheres.

Dominant as it was for centuries, even in antiquity the Aristotelian worldview was not without competition. Prior to Aristotle, the Greek philosopher Anaxagoras of Miletus had claimed that the universe is infinite, static, and homogeneously filled with matter. Imagine this matter to be shining stars, and you will have trouble understanding the darkness of night. Regardless of where we look, in such a universe our gaze will eventually meet the bright surface of a star (Figure 3.5). This is similar to the situation of a hiker in a forest. On looking around in any direction, his sight will eventually end at the trunk of a tree. Some trees are far away and will take up only a little of his field of view, others are nearby and obscure the view of more distant trees. The same would happen with stars if the universe had the aforementioned features: our sight would always end at a star's bright surface, and gleaming light would come from

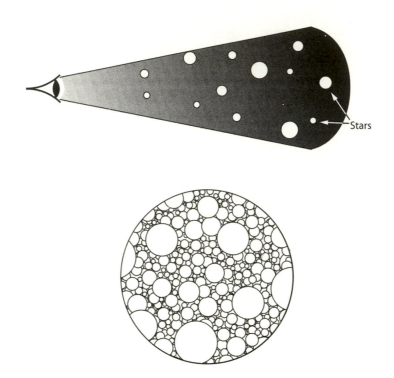

Figure 3.5 Not every cosmology is reconcilable with the darkness of night. In an infinitely large, static universe uniformly filled with luminous stars, twinkling starlight would come at Earth from every direction. Looking through a telescope, we would see the stellar disks overlapping each other. After Edward Harrison, *Darkness at Night: A Riddle of Cosmology* (Cambridge, Mass.: Harvard University Press, 1987), 7.

all directions in the sky. In the eighteenth and nineteenth centuries, the astronomers Jean-Philippe de Chéseaux and Wilhelm Olbers independently proposed that the presence of obscuring matter between the stars could save the cosmological model of Anaxagoras, for such matter would absorb most of the starlight. Today, however, we know that the laws of thermodynamics forbid such a solution. Once illuminated by all the stars around, the matter would start to gleam as brightly as the stars themselves. There is no escape from the conclusion that the darkness at night rules out a universe that is infinite, static, and homogeneously

Figure 3.6 In a clear night, the space between stars looks black. This photograph shows the Hyades (left) and the Pleiades (right), two open star clusters in the constellation Taurus. Courtesy of Sebastian Voltmer.

filled with stars. But this does not necessarily mean that the competing cosmological model of Aristotle is true.

Famous scholars have tackled the riddle of cosmic darkness (Figure 3.6), among them Johannes Kepler, Edmond Halley, and Lord Kelvin. Surprisingly, one of the first thinkers to come close to the current solution was not a professional scientist but a poet. Edgar Allen Poe, the author of famous works such as *The Raven* and *The Fall of the House of Usher*, gave a lecture at the Society Library in New York in 1848. Few listeners were present for his address, "The Cosmogony of the Universe." Imagining a cosmos that periodically contracts and expands with the pulse of the Divine Heart, Poe was not received very well by his scientifically minded contemporaries. However, by proposing a universe that was of a finite age and by being aware of the finite speed of light, Poe suggested two plausible reasons why the universe cannot be filled with light, and must therefore be dark.

Only since the 1960s do we have a theory that includes the two conditions proposed by Poe and is consistent with a vast number of astronomical observations—the Big Bang model. There is much converging evidence to support the idea that our universe formed about 14 billion years ago and has been expanding ever since. Therefore, the oldest existing stars may be a bit less than 14 billion years old, but most stars will be younger. Because of their finite number and lifetimes, as well as the limited speed of light, they cannot nearly fill the universe with light. The universe's expansion dilutes starlight even further. Thus, the universe is dark because there are not enough stars to fill it with light.

If Leonardo was right in claiming that the darkness behind the atmosphere must be involved in producing the blue of the daytime sky, then this color must be related to the history of the universe as a whole.

DISCOVERIES WITHOUT CONSEQUENCES?

Shortly after composing the *Codex Leicester*, Leonardo left Italy and emigrated to France, at the invitation of King François I. He settled in a mansion in Amboise, near one of the king's castles on the bank of the Loire River. In 1519 Leonardo died after suffering a stroke and was buried in a local cemetery.

In his studies, he had moved about freely, using a variety of texts from the optical tradition as sources of inspiration and reference. Spanning from Alberti back to Aristotle, his sources cover almost two millennia. An insatiable curiosity had made him a scholar who knew many of the texts that professional academics debated about. This was a remarkable achievement, given that he was able to read Latin texts only with considerable effort. Yet the astonishing depth and range of his studies stand in contrast to the meager influence that Leonardo achieved among scientists and engineers.

One reason for this may simply have been bad timing. Leonardo lived in a transition period between the late Middle Ages and the Renaissance. Academic life in those days too often consisted in the dry exegesis of dust-covered manuscripts of classical philosophers, a reading list partly censored by church officials. To be sure, the works of Aristotle were not regarded uncritically in fourteenth- and fifteenth-century Italy. Several precursors of what later became known as the scientific revolution of the seventeenth century can be traced back to the Renaissance period. But instead of moving forward to new concepts, Leonardo went back to the roots and turned to the Aristotelian tradition. In the end he remained a thinker who was in search of a system of thought.

Yet another reason for Leonardo's lack of influence may have been that he fell victim to his own perfectionism. Considering the wealth of detail in his observations and experiments, plus his tendency to accumulate ever more information, developing a synthesis of it all must have grown more and more difficult. When he died in 1519, he had not finished even one of the several books that he had announced. It became his pupils' duty to collect and publish at least his most important artistic advice as contained in the notebooks.

Francesco Melzi, who had accompanied Leonardo to France, inherited most of the notebooks and began to compile the *Treatise on Painting* (Trattato della Pittura), a book that summarized his ideas about the sky's blue. When the *Treatise*'s first printed version appeared in 1651, it had already been well-known among painters of the sixteenth and early seventeenth centuries. A number of handwritten copies had been circulating in the art world. In the late seventeenth century and throughout the eighteenth century, the *Treatise* became obligatory reading for artists. The book was translated into all major European languages. Not until the mid–nineteenth century were his manuscripts more thoroughly edited, and consequently the wide horizon of his thinking became evident. But in the meantime, most of his discoveries and inventions had been repeated and published by others.

CHAPTER 4

A Color of the First Order

The Great Plague hit Cambridge in the summer of 1665. By September, the city's government had forbidden all public meetings in town, and in October even the sermons in Great St. Mary's, the principal church, were canceled, for the danger of contagion was deemed too great. By then, all the colleges of the city's famous university had long been closed. Teachers and students had been urged to leave town and wait in the country for the plague to end. In comparison with earlier outbreaks in the Middle Ages, this time its devastation was delimited; still, public life in Britain ground to a halt, exacerbating a time of uncertainty. Just a year before, British troops had captured the Dutch settlement New Amsterdam on the American East Coast, renaming it New York. The Dutch reacted quickly to this affront, and soon the Second Dutch War broke out. It was to rage on for the next two years, culminating in the Great Fire of London in September 1666, which, according to various sources, was caused by arson.

When Trinity College in Cambridge was closed in August 1665, Isaac Newton (Figure 4.1) was among those sent home. The twenty-two-year-old student returned to his family's manor in Woolsthorpe, a small town in the East Midlands of England (Figure 4.2). The manor still exists today, and it seems to have changed little since then. Now a museum run by the British National Trust, it is surrounded by orchards, paddocks, farmhouses, and pastures full of grazing Lincoln Longwool sheep and hens. The two-story gray limestone building is furnished in the style of a prosperous yeoman's manor from Newton's time. The ground floor comprises the kitchen, the main hall, and the parlor. A staircase leads to two bedrooms on the upper floor. In some parts of the house graffiti can be seen on the walls, much of it probably written by Newton himself— some perhaps originating from the plague years. While waiting to return

Figure 4.1 Portrait of Isaac Newton in a steel engraving dating
from the mid-nineteenth century. Courtesy of Bildarchiv Preussischer Kulturbesitz, Berlin.

to Cambridge, Newton did not sit idly. Instead, he immersed himself in
studies of natural philosophy.

Newton was born in one of the two bedrooms on the upper floor of
Woolsthorpe Manor on Christmas Day, 1642. His father, a well-to-do
farmer, had died two months before his birth. When his mother remar-
ried in 1645, Newton was left in the care of his grandmother until his
mother, Hannah Newton, widowed for the second time in 1653, returned to Woolsthorpe. A certain Miss Storer, a girl somewhat younger

Figure 4.2 Woolsthorpe Manor near Grantham in Lincolnshire, in a woodcut from around 1890. Courtesy of Archiv für Kunst und Geschichte, Berlin.

than Newton, called the boy a "sober, silent, thinking lad," and it seems that she later became the only woman with whom he was ever romantically involved.[1] Newton's uncle recognized the growing boy's aptitudes and supported him. His schoolmaster is reported to have said that it would be a loss to bury so extraordinary a talent in rural pursuits, an enterprise that was bound to fail.[2] The nineteen-year-old Newton was sent to Cambridge to study at Trinity College.

Natural philosophy had progressed by leaps and bounds in the century and a half between the death of Leonardo da Vinci and Newton's arrival in Cambridge. Mighty upheavals in mechanics and astronomy had long since run their course. Nicolaus Copernicus' cosmology had banished Earth from the center of the universe and replaced it with the sun. Galileo Galilei and Johannes Kepler had shown that the old notions about the natural movements of the elements were no longer tenable. For many, these were good reasons to question the whole of Aristotelian natural philosophy. These were all new developments, and the chancellors of the university were at odds over how to design a new curriculum. This suited Newton perfectly, and he took advantage of the situation by freely sampling from the courses offered at his college.

While the new advances in mechanics came to be accepted within a short time, neither Copernicus nor Galileo nor Kepler contributed new ideas about the nature of light and color. The Copernican revolution was a revolution in mechanics, but not in optics. In the early seventeenth century, more and more scholars began to realize that a revolution was due in this field as well. Famous philosophers such as René Descartes and Francis Bacon, as well as Pierre Gassendi, were working on new theories of light and color that had little in common with that of Aristotle.

Cambridge University reopened in 1667, after the Great Plague ended, and Newton returned in the summer of that year. Within eighteen months he had not only gained new insights into the nature of light and color but had also discovered the law of gravitation and invented differential and integral calculus. The initial purpose of Newton's optical experiments was to test some contradictory theories of light. René Descartes, for example, imagined that light must consist of the smallest particles of matter, the *corpuscula*. The speed of their rotations would determine which color impression was formed in the eye. Descartes had predicted that, after refraction, light must move in arched trajectories. This was a statement that could be tested. By devising and conducting a suitable experiment, Newton became a representative of the experimental method that Galileo had founded half a century before. This entailed carrying out directed experiments in the controlled conditions of a laboratory in order to formulate empirical laws or to test hypotheses. The prism experiments Newton devised appeared to meet these criteria. Newton could confirm neither Descartes' hypothesis nor any of the other ideas, and soon reached the conclusion that a new theory of light was needed in order to create order out of the chaos of optical phenomena.

THE CELEBRATED PHENOMENA OF COLORS

In a letter to Henry Oldenburg, secretary of the Royal Society, dated February 6, 1672, Newton describes the first prism experiment he did on the upper floor of Woolsthorpe Manor:

> In the beginning of the year 1666 (at which time I applied myself to the grinding of Optick glasses of other figures than Spherical) I procured me a triangular glass Prism to try therewith the celebrated phenomena of Colours. And in order thereto having darkned my chamber and made a small hole in my window-shuts to let in a convenient quantity of the Sun's light, I placed my Prism at its entrance, that it might be thereby refracted to the opposite wall . . . applying myself to consider . . . [the colors] more circumspectly, I became surprised to see them in an oblong form which according to the received laws of refraction, I expected should have been circular . . . Comparing the length of this Coloured Spectrum with its breadth I found it about five times greater, a disproportion so extravagant that it excited me to a more than ordinary curiosity of examining from whence it might proceed.[3]

Figure 4.3 shows the design of the experiment. Newton allowed sunlight to shine onto a prism through a small, round opening in the window of a darkened chamber. The prism split the white sunlight into the colors of the rainbow and projected them onto the opposite wall of the room. There, the colors from violet to blue, cyan, green, yellow, orange, and red could be seen in an oblong array. Crucial to this experiment is the distance from the prism to the wall onto which the fragmented sunlight is projected. Only when the wall is quite far from the prism can the separation of the spectral colors be seen clearly. In his experiments, this distance equaled about seven meters.

By splitting up sunlight with a prism, Newton was investigating a phenomenon that had been known since the Middle Ages. It had been observed that, when lenses are used, colored fringes often appear. These

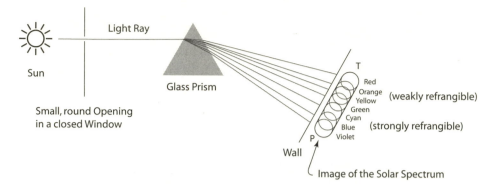

Figure 4.3 Isaac Newton used a glass prism to split white sunlight into the colors of the spectrum from violet to red. The oblong spectrum is made up of overlapping images of the solar disk, which are of different colors because they are refracted to different degrees.

phenomena were called "chromatic flaws" and were often treated as such, being considered undesirable aberrations. Newton, on the other hand, referred to them as "celebrated phenomena of colors," and followed the experiment described above with a series of further investigations. He concluded from his experiments that white sunlight is a mixture of many light rays in a variety of colors, which can be separated owing to a property he called refrangibility. Measuring the angles resulting from the refraction of different-colored rays of light he determined that

> This Image or Spectrum PT was coloured, being red at its least
> refracted end T, and violet at its most refracted end P, and yellow,
> green and blue in the intermediate Spaces. Which agreed with the
> first Proposition, that Lights which differ in Colour, do also differ
> in Refrangibility.[4]

Refrangibility thus indicates the degree of deflection of the light rays when they are refracted by the prism. Red light has the greatest refrangibility and violet light the smallest, while the refrangibilities of the yellow, green, and blue light rays fall in between. It is owing to the different refrangibilities of the light rays contained in sunlight that the prism is able to separate white light into the colors of the spectrum. Newton designated pure colors, which result from the refraction of light by a prism and cannot be broken down into other colors through the intercession of an additional prism, as "primary colors." Although he originally identified eleven such colors, in the *Opticks* he would later reduce their number to seven. In order of increasing refrangibility these are red, orange, yellow, green, blue, indigo, and violet. Newton emphasizes that the notion of primary colors is a simplification, because there must be infinitely many degrees of refrangibility, and that would correspond to infinitely many colors.

According to Newton, black and white are not colors, because black represents the absence of light, and in his first experiment he had revealed white to be a composite color. Newton's theory thus contradicts the natural philosophy of Aristotle; nonetheless, both the Greek philosopher and the British physicist believed that a harmony comparable to music could be found in the nature of light, and they associated the primary colors with the seven basic tones from A to G. Newton insists that we must distinguish between the refrangibility of light rays and the perceptions caused by them:

> And if at any time I speak of Light and Rays as coloured or endued with Colours, I would be understood to speak not philosophically and properly, but grossly, and accordingly to such Conceptions as vulgar People in seeing all these Experiments would be apt to frame. For the Rays to speak properly are not coloured. In them there is nothing else than a certain Power and Disposition to stir up a Sensation of this or that Colour.[5]

Thus, according to Newton, colors arise only in our perception; neither the objects seen nor the light rays are colored. Newton's world is colorless and is only perceived by us as colored because we translate the differing refrangibilities of light rays into the colors of the spectrum. So for Newton, the sky's blue originates in our perception. In this sense, the sky is not itself blue but rather sends rays down to us that are predominantly of high refrangibility and are translated by our perception into the color blue. However, what remained to be determined was a physical mechanism that could explain the splitting (dispersion) of light in the atmosphere, as well as the predominance of refrangible rays in the light from the sky. In what follows, I will continue to speak in terms of the blue color of the sky, as well as the other colors.

A NEW COLOR THEORY

Newton did not have to wait long for his achievements to be appreciated. His breakthrough work in optics and mathematics was recognized almost immediately upon his return to Cambridge. In the autumn of 1667 he became a fellow of Trinity College. Two years later, his teacher Isaac Barrow abdicated his chair so that Newton could take it over. Newton thus became the Lucasian Professor of Mathematics and Astronomy at the age of twenty-six. He held that honorable position for the next thirty-two years.

From his prism experiments Newton had concluded that all the colors of the spectrum must combine to produce white. However, in January 1673 he received a letter from Christiaan Huygens, a Dutch physicist, in which the latter related his observation that white light could be produced even from a mixture of yellow and blue rays only. Newton was flabbergasted and perceived this finding as a critique of his theory. At first he tried to refute Huygens by writing that the mixture of yellow and blue was "not completely" white, but he would never again claim that all the colors of the spectrum were needed to produce a white mixture.

Two decades later, while writing his book *Opticks*, Newton transformed Huygens' observation into a triumph for his own color theory. With the color circle, he presented a simple tool for predicting the hue and saturation that will result from the mixture of two different-colored light rays (Figures 4.4 and 4.5). This color circle is formed by arranging the seven basic colors in a circle, with magenta (a blue-tinged red not contained in the spectrum) inserted between violet and red. Remember that Newton perceived an analogy between colors and the musical scale.

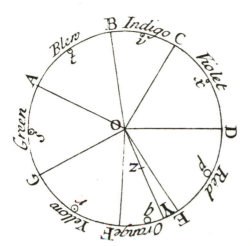
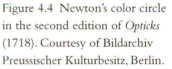
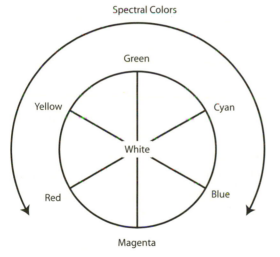

Figure 4.4 Newton's color circle in the second edition of *Opticks* (1718). Courtesy of Bildarchiv Preussischer Kulturbesitz, Berlin.

Figure 4.5 Simplified depiction of Newton's color circle.

Accordingly, the portions of the circle's arc that he assigns to the basic colors correspond to their position on the musical scale. Finally, the middle point of the portion corresponding to each color is marked along the circumference of the circle.

To determine what color will result when two different-colored light rays are mixed, you first connect the corresponding points along the

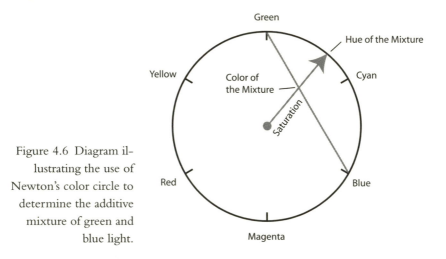

Figure 4.6 Diagram illustrating the use of Newton's color circle to determine the additive mixture of green and blue light.

circumference (Figure 4.6). When the intensity ratio of the two light rays is known, it is possible to pinpoint the exact location of the light mixture. From this ratio the position of the light mixture on the line segment between the two original colors on the color circle can be calculated. By projecting this point from the center of the circle outward, we can find the hue of the resulting light mixture on the circumference of the color circle. The saturation—that is, the purity of the color—corresponds to the distance from the center of the circle to the position of the color mixture. The colors along the circumference are fully saturated, while the center point of the circle is fully unsaturated. Here we find white, as the mixture of all the colors along the circumference.

This may sound quite abstract at first, but an example can illustrate the color circle's usefulness. Take Huygens' mixture of yellow and blue light, two colors that lie on opposite sides of the color circle. If both light rays have the same intensity, then the mixture falls in the center of the circle and is white. Huygens was correct in his observation. For another example, take the mixture of blue and green light, as shown in Figure 4.6. The result is cyan, a blue-tinged green. Likewise, Huygens' observation of white resulting from the mixture of two colored light rays can clearly be

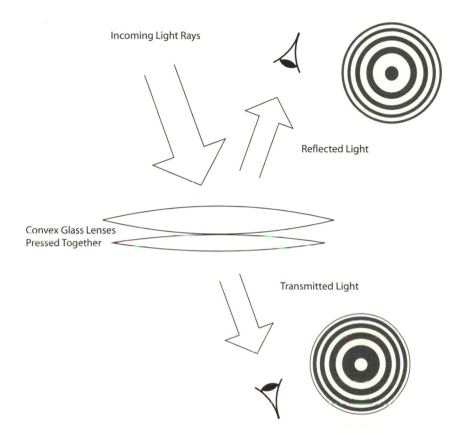

Figure 4.7 When two convex lenses are pressed together, Newton's colored rings can be seen in the reflected and transmitted light.

replicated with any pair of colors that are located opposite each other on the color circle. Pairs that enter into this relationship are called complementary colors. Yellow is thus the complementary color of blue and red is the complementary color of cyan.

All of the characteristics of color mixtures that we have dealt with here pertain only to the mixing of colored light rays. This is referred to as the additive mixing of color (Color Plate 12). It plays an important role not only in the sky, but also in color television and in stage lighting. In contrast, the mixing of pigments in painting is not additive but subtractive mixing, which follows different rules (Color Plate 13). Whereas the additive mixing of light rays of every color in the spectrum results in white light, the subtractive mixing of all the pigments on a painter's palette produces black.

COLORFUL RINGS

With his prism experiments, Newton had investigated only the colors of light. He now faced the challenge of finding an explanation for the colors of material objects as well. He was probably already engaged with this problem when he conducted his prism experiments at Woolsthorpe. He took as a point of departure a discovery that Robert Hooke had made with mica a few years before. In its natural state, the mica appeared colorless. But when it was cut into thin sheets, a shimmering array of colors appeared. This surprising finding prompted Hooke to carry out a thorough investigation, which led to the conclusion that the apparent colors of the sheets depended on their thickness.

Newton discovered similar color phenomena when he pressed two convex lenses together and illuminated them with muted sunlight (Figure 4.7). In both the light shining through the lenses and in the reflected light he saw a series of concentric rings whose colors resembled those of the mica sheets. These rings have been known ever since as Newton's rings (Figure 4.8). They remind one of the colors visible in soap films (Color Plate 14). On careful inspection, Newton determined that, within the sequence of the rings, the colors of the spectrum are more or less repeated, and he assigned these colors to groups that he called "orders." When he pressed the lenses together very firmly and observed the reflected light, he saw the following colors, from the inside progressing outward. The letters in parentheses refer to Newton's drawing reproduced in Figure 4.8:

First Order: black *(a)*, blue *(b)*, white *(c)*, yellow *(d)*, red *(e)*.
Second Order: violet *(f)*, blue *(g)*, green *(h)*, yellow *(i)*, red *(k)*.
Third Order: purple *(l)*, blue *(m)*, green *(n)*, yellow *(o)*, red *(p)*.
Fourth Order: green *(q)*, red *(r)*.[6]

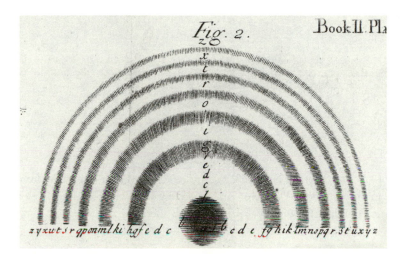

Figure 4.8 Newton's rings in the *Opticks*. One half of the symmetrical pattern of rings is shown here. Dark segments are shaded. The sequence of colors is labeled with the letters *a* to *z* (see text). The blue of the first order makes up the innermost colored ring (*b*). Courtesy of Bildarchiv Preussischer Kulturbesitz, Berlin.

Beyond the seventh order he could no longer make out the colors, because the distance between the rings becomes smaller and smaller there, and the colors overlap more and more. The colors in the reflected light are different from those seen in the transmitted light. In every spot where a color appears on one side, its respective complementary color is seen on the other side (Figure 4.9).

In other experiments Newton illuminated the stacked lenses with monochromatic light, which he had first separated out of the white sunlight with a prism. Now the rings exhibited only the color of the illuminating light, but they were more clearly discernible as a sequence of lighter and darker segments. The especially refrangible violet light produced the smallest rings, while the red rings were the largest.

Once Newton had identified these regularities of the colors in the rings, he began to search for their cause. It was clear that the space between the convex lenses increased with the distance from their point of contact, and it seemed to him that the size of this gap must determine

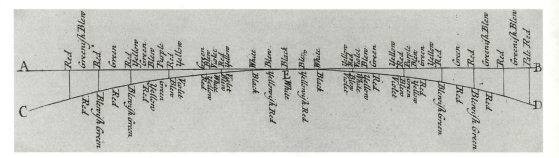

Figure 4.9 The colors of Newton's rings in the reflected light (top) differ from those in the transmitted light (bottom) and form pairs of complementary colors. Courtesy of Bildarchiv Preussischer Kulturbesitz, Berlin.

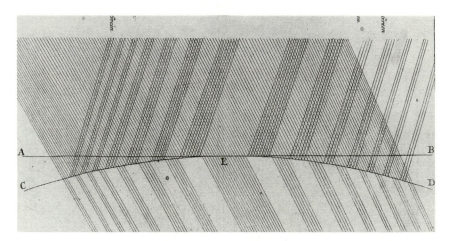

Figure 4.10 The colored ring phenomena forced Newton to revise his views on light. He now posited that light is periodic in nature, claiming that it possesses "fits of reflection" and "fits of transmission." The sequence of these fits was supposed to explain Newton's rings in the reflected and transmitted light. Courtesy of Bildarchiv Preussischer Kulturbesitz, Berlin.

whether the light would be reflected or transmitted at a given point (Figure 4.10). Furthermore it seemed as if the points of reflection and transmission depended on the refrangibility of the light rays. Newton knew the focal distances of the stacked lenses—and hence the shape of their curvature—and so he was able to calculate the distance between the lenses for each given point. In addition, he measured the colored rings' diameters and could thus assign all the colors to the corresponding sizes of the gap. It turned out that light of a certain color is reflected when the size of the gap is an integer multiple of some small distance. The sizes of the gaps at which light is transmitted through the lenses lie precisely between these values. With illumination by white sunlight, the gap at the first dark ring, the one between the colors of the first and second orders, measures 1/89,000 inch, according to Newton. That is about 1/3,500 of a millimeter, the smallest distance measured up until that time.

But what lay at the root of these stupefying regularities? Newton had to rack his brain to interpret the colored rings. Up until then he had assumed that a light ray consisted of an irregular sequence of small particles that he called "corpuscles." These particles, Newton thought, traveled in straight lines and were reflected and refracted according to certain laws. When a composite light ray is split due to refraction by a prism, he wrote, we see how the corpuscles of different-colored light rays are refracted to varying degrees, according to their refrangibility.

Even before his investigation into the colored rings, Newton was aware of a weak point in this simple corpuscle model. It was common knowledge that, when a light ray strikes a pane of glass at an oblique angle, part of it is reflected and the remainder is refracted through the glass. So wouldn't there have to be at least two kinds of corpuscles, differing from each other in that some are reflected while others are refracted? Although this problem did not concern Newton too much at first, the ring phenomena forced him to modify his conceptions of light. In *Opticks* he writes that light must have an intrinsic periodic quality that determines whether it will be reflected or transmitted. He calls this quality "fits" of reflection and transmission. According to Newton, the length of these fits is related to the refrangibility of the light rays. Highly refrangible blue light has the shortest fits, and red light, being the least refrangible, has the longest. Perhaps, he thought, periodic fits excited waves in the aether that were periodic as well.

Bear in mind that Newton's objective in investigating the ring phenomena was actually to explain the colors of objects. The most promising indication that he would succeed in this was the relationship between the color of the reflected or transmitted light and the size of the gap be-

tween the two lenses. Newton linked this observation to a hypothesis
about the composition of matter and wrote that all substances consisted
of the smallest particles, whose size varied according to the type of material. In every case, these particles were transparent and could reflect, re-
fract, or transmit light. Therefore it depended on the size of the particles
whether light of a specific refrangibility would be reflected or allowed
to pass through:

> The transparent parts of Bodies, according to their several sizes,
> reflect Rays of one Colour, and transmit those of another, on the
> same grounds that thin Plates or Bubbles do reflect or transmit
> those Rays. And this I take to be the ground of all their Colours.[7]

Attributing the variety of colors exhibited by objects solely to the size
of their constituent microscopic bodies, which are colorless in and of
themselves, may seem like a bold claim. But as a scientist seeking to reduce the diversity of natural phenomena to just a few causes, Newton
must have found this explanation immensely satisfying.

The innermost ring is an extremely weak blue, which for Newton
was the color of the lowest degree of refrangibility, except for violet.
He calls it the "blue of the first order." Wherever this color is found in
nature, it must correspond to the smallest particles capable of producing
color. As a familiar example, he cites the blue of the sky:

> The blue of the first Order, though very faint and little, may pos-
> sibly be the Colour of some Substances; and particularly the azure
> Colour of the Skies seems to be of this Order. For all Vapours
> when they begin to condense and coalesce into small Parcels, be-
> come first of that Bigness, whereby such an Azure must be re-
> flected before they can constitute Clouds of other Colours. And
> so this being the first Colour which Vapors begin to reflect, it
> ought to be the Colour of the finest and most transparent Skies,
> in which Vapors are not arrived to that Grossness requisite to re-
> flect other Colours, as we find it is by Experience.[8]

Newton thus recognized the blue of the first order in the blue color of
the sky. For him, this was proof of the extremely small size of the vapor

particles in the atmosphere. This finding was a note that Newton made in passing, and he went on to explain the colors of foam, paper, gold, and copper.

After his initial studies of the colors caused by refraction and reflection, Newton returned to the field of optics from time to time when he was forced to revise his concept of light. One persistent puzzle was the apparent periodicity inherent in light that led him to develop the theory of fits. As the historian of science Alan Shapiro has demonstrated with a meticulous examination of Newton's manuscripts, this would prove a vexingly difficult endeavor for the great scholar. Only in the early 1690s did Newton feel ready to assemble his findings into a book on optics. And another twelve years passed on top of that before the *Opticks* was finally published, in 1704 (Figure 4.11). Why did it take him so many years to publish his intriguing findings? Shapiro concludes that Newton had aimed at making the *Opticks* a complete treatment of the subject, but along the way he stumbled onto another problem: the "inflexion" of light. The rectilinear propagation of the light's corpuscles would lead one to expect a single, sharp division between darkness and light at the edge of a shadow. However, the Italian scholar Francisco Maria Grimaldi had noticed in the 1660s that instead of this sharp division there were bands of light and darkness running parallel to the expected shadow line. These were seen only in shadows cast by narrow beams of light. Though he was probably unaware of Grimaldi's discovery, Newton knew of this phenomenon, today known as diffraction, by 1675. The difficulty of adequately accounting for diffraction caused a five- to six-year delay in the publication of his book. In the end he had given up, suspecting that light is repelled somehow when passing near the corners of solid objects.[9]

There were more mundane reasons for the delay in finishing *Opticks* as well. From the mid-1690s, Newton became more and more deeply involved in British public life, leaving him less time for scientific work. In 1696 he was named Warden of the Royal Mint in London, and was promoted to the mint's Master four years later. He held this position until his death. Newton seems to have devoted much of his energy to this job, which included implementing a recoinage to stabilize the currency and prosecuting counterfeiters. In 1703, he resigned the Lucasian chair

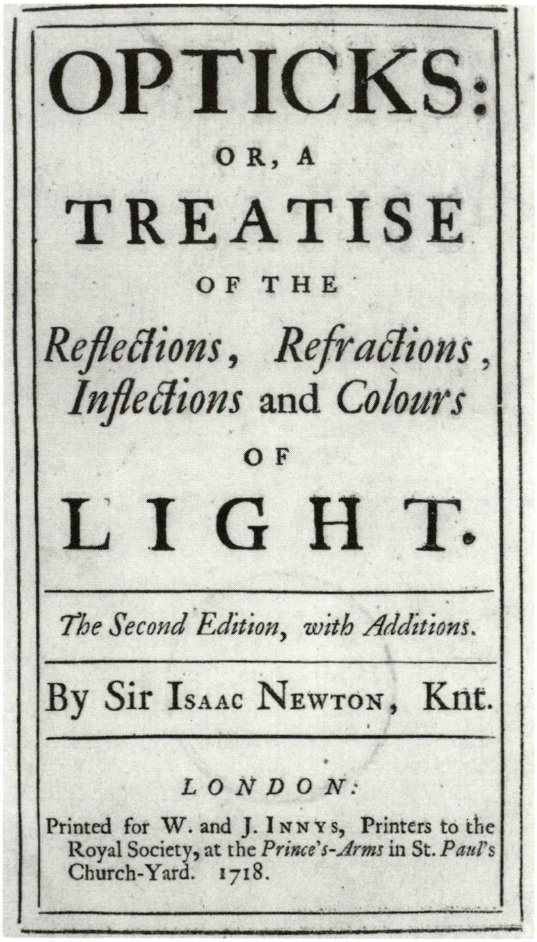

OPTICKS:

OR, A

TREATISE

OF THE

Reflections, Refractions,
Inflections and *Colours*

OF

LIGHT.

The Second Edition, with Additions.

By Sir Isaac Newton, Knt.

LONDON:

Printed for W. and J. Innys, Printers to the
Royal Society, at the *Prince's-Arms* in St. *Paul's*
Church-Yard. 1718.

Figure 4.11 Title page of the second edition of Newton's *Opticks* (1718). When the first edition was printed in 1704, Newton's name did not appear on the title page. Historian of science I. Bernard Cohen speculated that this was Newton's way of paying homage to Christiaan Huygens, whose *Traité de la Lumière* (1690) was likewise published without the author's name. For contemporary readers there was, however, no question as to the identity of either book's famous author. By writing in vernacular English and French rather than in scholarly Latin, both scientists drew attention to the preliminary nature of their optical investigations. Courtesy of Bildarchiv Preussischer Kulturbesitz, Berlin.

in Cambridge and became the president of the Royal Society. In the last decades of his life, the "sober, silent, thinking lad" had been transformed into a powerful personage in British public life, and he used this position to advance his own interests. Not mellowing in his old age, Newton engaged in a dispute of priority with Gottfried Wilhelm Leibniz over the invention of differential and integral calculus, a conflict that lingered on for years. In 1727, at the age of eighty-five, Isaac Newton died in London. He was buried in Westminster Abbey, next to statesmen and kings.

DISCOVERING THE ATMOSPHERE

In writing about the condensing vapors in the air, Newton was alluding to the book *The Sceptical Chymist*, published by his contemporary Robert Boyle in 1661. In that work, Boyle had characterized atmospheric air as a mixture of the "actual" air and foreign water vapor that arises from the earth. But Boyle vehemently rejected the notion that air was one of the four classical elements. The vagueness with which Newton seems to refer to Boyle's conception of air hints at its uncertain ontological status in seventeenth-century natural philosophy. Indeed, while the concepts that Newton and his colleagues had developed about the nature of light and color were soon praised as scientific progress, the microscopic structure and composition of air remained a mystery.

Since the late 1670s, when Newton had pursued laboratory studies in alchemy, he had been corresponding with Boyle about air. In 1679, Newton began to compose a treatise of two chapters, *Concerning the Air and the Aether*. Chapter 1 deals with the mechanical properties of air, in particular its expansion when heated. He speculates that air is made up of small particles imbued with mutually repulsive forces:

> But as it is equally true that air avoids bodies, and bodies repel each other mutually, I seem to gather rightly from this that air is composed of the particles of bodies torn away from contact, and repelling each other with a certain large force.[10]

Newton relates the macroscopic properties of air to its constituent microscopic particles. In Chapter 2, he continues this line of thought by contemplating the origin of the aether. Whereas Aristotle had considered this to be the element of the perfect celestial spheres, Newton wonders whether all space may be filled with this substance. He imagines that the aether may be generated from further fragmentation of the aerial particles into even smaller particles. But then, in the middle of a sentence, he abandoned his text, and never resumed writing it. It remained unpublished throughout his lifetime.

What had happened? Newton's biographer Richard Westfall suspects that Newton may have been conducting some new experiments that suddenly led him to conclude that the aether might not exist. Only much later, in a Query added to his *Opticks* of 1704, did Newton return to the conviction that the aether did exist after all. He now considered it necessary for the transmission of forces acting at a distance, such as gravity.

Not studying such forces himself, Boyle was less concerned with the aether than Newton and distinguished it sharply from air. In his *General History of the Air*, published posthumously in 1692, Boyle defines air thus:

> By the *Air* I commonly understand that thin, fluid, diaphanous,
> compressible and dilatable Body in which we breath, and wherein
> we move, which envelops the Earth on all sides to a great height
> above the highest Mountains; but yet is so different from the *Aether*
> [or *Vacuum*] in the intermundane or interplanetary Spaces, that
> it refracts the Rays of the Moon and other remoter Luminaries.[11]

Boyle characterizes air by both its mechanical and its optical properties, and explicitly stresses that, by "air," he does not mean one of the four classical elements but rather atmospheric air. This, he says, consists of three types of particles: first, the dry exhalations of the earth; second, very small particles radiated from the sun and other stars, which can luminesce on their own; and finally, a third kind of particle whose nature is supposed to explain the permanent elasticity of air. This latter is known to us from the way pneumatic bicycle or car tires remain inflated for long periods of time. Boyle suggests we could picture this last type as small feathers, filaments, or sawdust. As such, he distinguishes the properties of air from those of its constituent particles. There is, he admits, a wide array of other hypotheses that one could entertain but that could not be proved.

What is certain is that Boyle considered air to be a compound mixture situated in a space between the surface of Earth and a certain upper limit. But within this space, does it have a vertical structure of varying density or pressure? The Aristotelians had argued that the air we breathe is situated in its natural place, the sphere of air. If so, they concluded, it must be weightless, for if it had weight, this would indicate its desire to move away from its natural place—an obvious contradiction. Consequently, there could be no changes in air pressure or density with increasing altitude. Similarly, the Aristotelians considered the water in the ocean to be weightless, for it was regarded as being situated in its proper, natural place as well.

In the seventeenth century, scientists began to question the notion of the air's "natural place" by using the air pump to measure its weight. Soon after the invention of the suction pump, it was determined that it could raise water no higher than about thirty-three feet. This was blamed either on technical defects in the pumps or on the porosity of the tubes then in use. Another possibility was that nature abhors a vacuum. Philosophers had debated this idea vigorously since antiquity, and Aristotle had been among those insisting that a vacuum could not exist anywhere in the universe. His followers claimed that all space is filled with some kind of matter. In this view, the water's rising when the suction pump was used could be seen as its intentional action to prevent a vacuum from forming in the tube. Some seventeenth-century experimenters considered the maximum height of a column of water raised by a pump to be a measure of nature's abhorrence of a vacuum, the so-called *horror vacui*. However, the idea of nature abhorring a vacuum, and thus acting intentionally, fell more and more out of tune in an era increasingly dominated by the new mechanistic philosophy inaugurated by Galileo Galilei. In 1644, the Italian mathematician Evangelista Torricelli took a fresh look at the problem, abandoning the notion of "abhor-

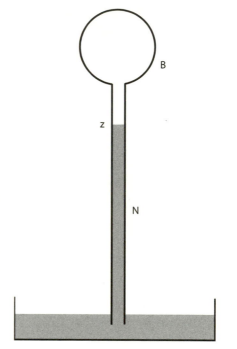

Figure 4.12 Sketch of Torricelli's
barometer. It consisted of a glass bulb
(B) with a long neck (N). In his experiment, Torricelli filled the bulb and the
neck with mercury and then inverted it
in a basin that contained more mercury.
The mercury column fell to a certain
level (z), and then rose or sank according to the atmospheric air pressure.

rence" and instead comparing the experimental setup with a mechanical
balance. Torricelli noticed that the air pump generated a column of water inside a vertical tube, while outside there was a column of atmospheric air. If the atmospheric air did possess weight, then the weight of
the water inside the tube could be considered the equivalent of the
weight of atmospheric air above the tube.

Torricelli proposed an experiment to test this hypothesis. He suggested sealing a glass tube on one side, filling it with mercury, inverting
it, and lowering it into a basin of mercury (Figure 4.12). Knowing that
the density of mercury is fourteen times that of water, he predicted that
the mercury column would come to rest at one-fourteenth the height
of the water column, or twenty-nine inches, because then the column
of water and the column of mercury would be of equal weight. When
the experiment was actually conducted, the height of the mercury column was measured at precisely the level that Torricelli had anticipated.

Torricelli drew a bold conclusion from this observation, writing: "We live at the bottom of an ocean of the element air, which by unquestioned experience is known to have weight."[12]

Torricelli's experimental setup of a mercury column in a glass tube reminds one of a barometer, and indeed it was the first instrument of this kind. However, his interpretation of the barometer as a balance to measure the weight of the air was not accepted unquestioningly. In France, the philosopher Blaise Pascal objected that by replacing the water in the tube with mercury, Torricelli had varied only one side of the balance. One would need to vary the other side as well, Pascal claimed, to make sure that the equivalence between the barometer and the mechanical balance was universally valid and to rule out the explanation based on nature's abhorrence of a vacuum. Pascal proposed that the other side could be varied by taking Torricelli's barometer to the top of a mountain and comparing its level with that measured in the plains below. If Torricelli was right, Pascal argued, the weight of the air above the barometer should be smaller when it was placed on top of a mountain, and thus the height of the column of mercury or water should be less than in the plains below.

The day of truth came in September 1648. Florin Périer, Pascal's brother-in-law, climbed to the summit of Puy de Dôme, a mountain in central France, and recorded the level of the barometer. Meanwhile a monk at a monastery recorded the level of another barometer that remained at the base of the mountain. When Périer returned, a comparison of the two barometer readings showed that on the summit, approximately 1,000 meters above the convent, the column of mercury had been about nine centimeters smaller. As Torricelli had predicted, the lesser amount of atmospheric air at the higher altitude imposed less weight on the barometer. This finding convinced Pascal that Torricelli was right in arguing that air has weight, the barometer is comparable to a mechanical balance, and a vacuum is possible in nature. Acting as a balance, the barometer reading showed that the weight of the 1,000-meter column of air is equivalent to that of a nine-centimeter column of mercury. The experiment thus revealed that both the pressure and the density of atmospheric air increase toward the ground because of the combined pressure—the weight—of the layers above. There are two further implications, one practical and the other cosmological. On the practical side, it was soon realized that the barometer recorded changes in air pressure that may be linked to changing weather conditions. In terms of cosmology, it led Pascal to the conclusion that perhaps even most of the universe's volume may be a vacuum, a space entirely devoid of matter.

Torricelli's finding that air has weight and Pascal's insight that the pressure of the air decreases with altitude are two pillars of atmospheric physics to this day. They make us wonder if there is an upper limit to the atmosphere. Five centuries prior to Torricelli and Pascal, the Andalusian mathematician Abu'Abd Allah Muhammad ibn Muadh had suggested that simple observations could provide an answer. In a short treatise later known as *On the Morning and Evening Twilight* (De crepusculis matutino et vespertino), this eleventh-century scholar argued that the angle of depression of the sun at the end of the evening twilight (or the beginning of the morning twilight) is related to the height of the "atmospheric moisture" responsible for the reflection of the sun's rays. Based on the duration of twilight, he estimated this angle at 18 degrees—in remarkable agreement with today's definition of astronomical twilight, that is, the beginning (or the end) of dark night. Ibn Muadh's reasoning is elaborated in Appendix A. From Ibn Muadh's 18 degrees and the radius of Earth as we know it today, we can infer the height of the atmosphere to be about 79.5 kilometers (50 miles). This is just over one percent of Earth's radius; the atmosphere is nothing more than a thin envelope around our planet.

A SKY FULL OF DROPLETS?

Newton's explanation of the sky's color as a blue of the first order resulting from tiny water droplets may seem quite plausible at first, but it was soon met with criticism. One challenge came from Johann Caspar Funck, from Ulm in the German province of Swabia. In his *Book on the Colors of the Sky* (Liber de coloribus coeli), published in 1716, this country

preacher took the world-famous physicist to task.[13] Funck observed
how the colors of the sky and the clouds change over the course of a
day: when the sun is low in the sky, reddish colors predominate, while
at noon the clouds look white and the clear sky looks blue. According
to Newton's theory, the cloud droplets would therefore have to be bigger in the morning and evening to be able to reflect the reddish light
rays. Conversely, around noontime they would have to come in a range
of sizes in order to reflect all the colors of the spectrum equally. Only
then could the white composite light of the clouds emerge. Yet the water droplets that are suspended in the air outside the clouds—and that,
according to Newton, make the sky look blue—could not change in
size, for then another color would appear. Funck criticized Newton for
not giving any plausible reason why the water droplets should change
in size. The Scottish physicist Thomas Melvill argued likewise, adding
that Newton's theory was hard pressed to explain the colors of clouds
at sunset:

> Why the particles of the clouds become just at that particular time,
> and never at any other, of such magnitude as to separate these
> colours; and why they are rarely, if ever, seen tinctured with blue
> and green, as well as red, orange, and yellow?[14]

But the size distribution of atmospheric water droplets was not the
only dilemma. Newton's explanation presupposes a perfect equilibrium
in the atmosphere whenever the sky appears blue. In other words, the
smallest, newly formed vapor particles would have to be present in
every part of the air at such times. Moreover, they would have to be
evenly distributed all throughout the air in order to produce the even
coloring we see. Yet although the air is subject to very large fluctuations
in humidity—depending on weather conditions—the sky's blue is strik-
ingly constant. On dry, hot summer days, the humidity is exceptionally
low; nonetheless, the sky is usually blue. Perhaps Newton would have
countered that the water droplets must be located higher up in the atmosphere at such times. In the late eighteenth century, the Swiss naturalist Horace Bénédict de Saussure constructed the first hygrometer and
used it to show that the humidity in the mountains does not rise with

increasing altitude but rather declines considerably. Newton's theory was running up against bigger and bigger challenges.

In the middle of the nineteenth century, two additional problems arose. In 1848, the Austrian physiologist Ernst Brücke studied the exact sequence of colors in Newton's rings and saw the blue of the first order as a lavender-gray hue. In his estimation, the saturation of this first color ring was much lower than that of the sky's blue, at least when clouds are absent.

What is more, Newton's explanation has a peculiar consequence. If there are many water droplets in the air, then the sun, moon, and stars should not appear as sharply defined objects. The myriad reflections in the water droplets would distort the images of the heavenly bodies into blurry, oversized disks of light (Figure 4.13). Rudolf Clausius, a physicist from Berlin, pointed this out in 1849 and tried to salvage Newton's theory by suggesting that it was not water *droplets* that were suspended in the air but rather tiny water *bubbles*. The presence of such bubbles would make the heavenly bodies appear with their familiar sharp outlines and would give the sky its blue of the first order. On the other hand, owing to capillary forces, these bubbles would quickly pop. No wonder that nobody could find them in the air.

It was not only Newton's explanation of the blue sky that provoked objections; even his theory of the colors of material objects came under fire, particularly in early eighteenth-century France. One of the most outspoken critics was Jean-Henri Hassenfratz, a master carpenter who eventually became a physics professor at the renowned École Polytechnique in Paris. Studying colored fluids through his prism, Hassenfratz grew more and more convinced that their apparent colors are caused by selective absorption of specific spectral colors rather than by the sizes of the smallest constituent particles. Ironically, it was with this very same prism that Hassenfratz gathered evidence supporting an essentially Newtonian framework for explaining blue skylight. In January 1801, he repeated Newton's experiment from the year 1666, breaking up white sunlight into the colors of the spectrum.[15] Hassenfratz carried out this experiment at different times of day, with the sun at various heights, and measured the length of the solar spectrum, noting the colors it contained

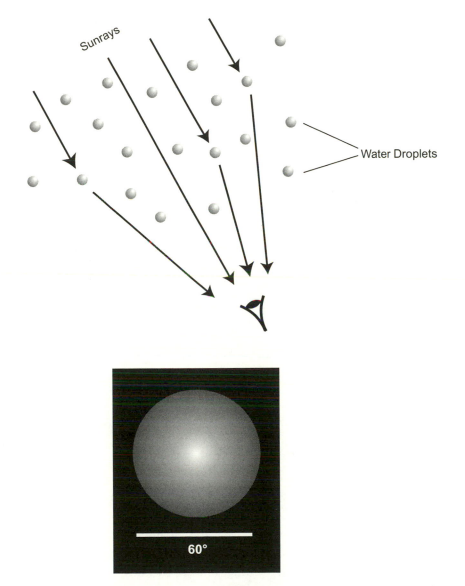

Figure 4.13 In 1853, the physicist Rudolf Clausius discovered a flaw in Newton's explanation of blue skylight: if it were caused by reflections off small water droplets, then the sun would have to appear as a large, blurry disk with a diameter of 60 degrees. But in fact, the visible solar disk is just one-half of a degree in diameter, and it has a distinct edge.

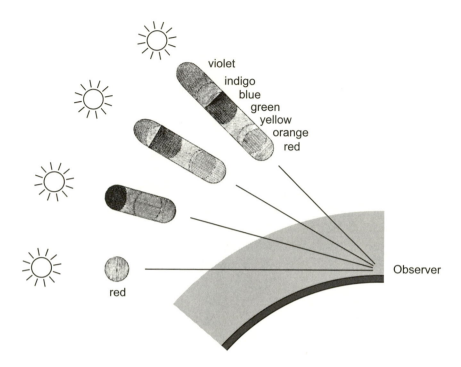

violet
indigo
blue
green
yellow
orange
red

Observer

red

Figure 4.14 On January 13, 1801, Jean-Henri Hassenfratz measured the apparent length of the solar spectrum over the course of a day. While the spectrum of the setting sun appeared as an orange circular disk, the oblong solar spectrum at noon—when the sun was at its highest—contained all the colors of the spectrum. It was thus proved that the atmosphere separates the colors of sunlight.

(Figure 4.14). He determined that at noon, when the sun is highest in the sky, sunlight contains all the colors of the spectrum, from violet to red. But when the sun is low in the sky, the violet, blue, and green rays are missing; the spectrum is distinctly shorter. This is in accord with our everyday experience that the setting sun appears red, while it looks whitish when it is higher up in the sky.

Since the path of the sun's rays through the atmosphere is much longer when the sun is low in the sky than at noon, in this case the atmosphere must have removed the refrangible violet, blue, and green light rays from direct sunlight. Only the less refrangible yellow, orange, and red light rays pass unhindered through the thick layers of air and make the rising and setting sun look orange-reddish. But where do the refrangible rays go when they are separated out of the direct sunlight? According to Newton's color circle, an additive mixture of violet, blue, and green (the colors that are missing from the spectrum at sunrise and sunset) yields a blue hue. Hassenfratz deduced that the refrangible light rays that are filtered out of the direct sunlight by the atmosphere light up the sky and make it appear blue.

COLORED AIR AND THE INHABITANTS OF THE MOON

Although numerous shortcomings had been found in Newton's microscopic analysis of the dispersion of colored sunlight, its basic tenet—the air's capacity to split white sunlight into rays of varying refrangibility, and thus of different colors—stood firm in the early nineteenth century. Most contemporary accounts just assumed that air or the particles contained in it reflect the refrangible (blue) light rays more than the less refrangible (red) rays, and left it at that. A look at physics textbooks of the eighteenth century shows how much uncertainty remained concerning this issue. In 1788, the German naturalist Friedrich Albrecht Carl Gren wrote in his popular *Foundations of Natural Philosophy* (Grundriss der Naturlehre) that the atmosphere reflects mainly blue light. Johann Heinrich Lambert, a physicist from Berlin, had taken a similar stance in the year 1774:

> The blue color that we see in very distant mountains is caused by
> the sunrays that are reflected in the air. The blue color of the sky
> can be traced back to these rays, because without these rays the
> sky would appear as dark [during the day] as it does at night.[16]

In Paris, the astronomer Pierre Bouguer claimed that the red light rays were more powerful than the blue ones and so could penetrate the atmosphere more easily. That explained why direct sunlight should appear yellow or reddish, while the weaker blue light of the sun gets stuck in the air, so to speak, and lends the sky its color. Bouguer thus sought the cause of the color splitting in the characteristics of light rays more than in the atmospheric particles. We have Jean-Antoine Nollet to thank for one fanciful depiction of this effect. This French priest and naturalist was a popular figure in his day, being widely known in French aristocratic and bourgeois circles for his public demonstrations of physics. Nollet's fame landed him a position as the physics teacher to the king of Sardinia, and eventually the Physics Teacher to the Royal Children of Louis XV, king of France. In his *Lectures on Experimental Physics* (Leçons de Physique Expérimentale), Nollet wrote that the sun's rays first pass through the atmosphere and are then reflected off the surface of Earth. Although the red rays are strong enough to traverse Earth's atmosphere a second time, he considered the blue rays to be weaker. These must get stuck in the air and give the sky its blue color. Consequently, writes Nollet, "the inhabitants of the moon, if there are any, would see our earth with its atmosphere as whitish and approaching blue."[17]

Not all eighteenth-century researchers referred to Newtonian optics in their musings on the sky's blue. Johann Christian Polykarp Erxleben of Leipzig was at a loss: "some naturalists affirm and others deny that air really is blue."[18] The best-known proponent of this view was the Swiss physicist and mathematician Leonhard Euler. Although air particles are only weakly bluish, he said, a large number of them could indeed yield an intense blue hue:

> We therefore do not detect any trace of this blue in a room; however, when all of the bluish rays from the entire atmosphere enter our eyes at the same time, even though the color of each individual ray is so weak, together they can bring about a strong and deep color.[19]

At the time Euler was a guest at the court of the Prussian kings, residing in Potsdam outside Berlin. He supported his claim with evidence from the forests of the Harz Mountains, a small range in northern Germany, about 120 miles west of Berlin. From the city of Halberstadt, on the edge of the Harz, the forests looked green. But from Magdeburg, forty miles farther away, they appeared bluish. Without realizing it, Euler had repeated an observation that had prompted Leonardo da Vinci to formulate the aerial perspective for painting. While Leonardo had at first also assumed a blue color for air, he nonetheless recognized the contradiction between this assumption and other observations. Euler, on the other hand, succumbed to the error of underestimating the complexity of the problem.

In his 1799 *Essay on Heat, Light and the Combinations of Light*,[20] the English chemist Humphry Davy claims that there are attractive and repulsive forces intrinsic to all bodies. Thus, solid objects are held together by the predominance of attractive forces, while in liquids the attractive and repulsive forces are in equilibrium. In light, on the other hand, the repulsive forces outweigh the attractive ones by far. That, Davy writes, is why it is so fast. Yet the strength of repulsion differs from one light ray to the next, and that is the reason for the diversity of colors. Red light is characterized by especially strong repulsive forces. Along the path it travels through the air, the repulsive force of a red light ray is diminished, says Davy, and makes it turn blue.[21] What may at first seem like a quixotic claim is rather inspired by mechanistic theories of light such as the one proposed by René Descartes. Eighteenth-century scholars were far from reaching a consensus on how to explain this color.

CHAPTER 5

Basic Phenomenon, or Optical Illusion?

On June 20, 1802, two men left Quito and headed for the summit of the world. They were out to climb Chimborazo, a steep pyramid of rocks and ice reaching high into the realm of eternal snow. Located south of the capital of a Spanish colony in what is today Ecuador, the mountain towered menacingly over the thinly populated plateau of Tapia. The two men were determined to reach the top, a feat that nobody had achieved before. Their names were Alexander von Humboldt and Aimé Bonpland. For three years, the German naturalist and the French botanist had formed a team of intrepid scientific explorers on an epic journey through Central and Latin America. That part of the world was largely unknown, and hardly imaginable, to scholars back home. Humboldt and Bonpland intended to introduce it to them by mapping the landscape, by gathering an immense collection of plant and animal specimens, and by taking measurements of Earth's magnetic field, the temperature and pressure of the air—and the blueness of the sky above. Having obtained a permit while in Madrid to conduct a scientific expedition in the Spanish colonies, they began their journey proper in the Spanish port of La Coruña. Here they boarded a ship to cross the Atlantic Ocean. After a brief stop in the Canary Islands, they arrived in Caracas, whence they explored the Orinoco basin, a large river system in Venezuela's tropical forests. This was followed by a tour of Cuba, after which they went south again to Colombia and the Andes.

Chimborazo had been considered the highest mountain in the world since 1746. In that year two Frenchmen, Pierre Bouguer and Charles-Marie de la Condamine, had measured the altitudes of Andean peaks and found Chimborazo to be the tallest of all. Bouguer and de la Condamine were the first to attempt the record-breaking climb to its summit, but,

Figure 5.1 Chimborazo, soaring over the plains of Tapia. Lithograph from a sketch by Alexander von Humboldt. Reproduced from Alexander von Humboldt and Aimé Bonpland, *Vues des Cordillères, et Monumens des Peuples de l'Amérique: Relation Historique, Atlas Pittoresque* (Paris: F. Schoell, 1810), Plate XXV.

poorly equipped and inexperienced in mountaineering, they soon gave up. Half a century later, Humboldt and Bonpland were better prepared. On their way from Europe three years earlier, they had climbed the 3,700 meters (12,140 feet) to the summit of Pico de Teide, the tallest mountain in the Canary Islands. After arriving in Quito in March 1802, Humboldt and Bonpland had refreshed their skills by climbing the slopes of several volcanic peaks nearby, reaching 4,411 meters (14,470 feet) at Cotopaxi and 5,405 meters (17,730 feet) at Antisana. They may have been physically fit, but their clothing was far from ideal. The two Europeans wore leather shoes and frock coats. They lacked gloves.

On June 23, three days after leaving Quito, the two men began their ascent proper at Calpi, a village at an elevation of 3,150 meters (10,335 feet). They had stayed overnight at the priest's house and packed their mule early in the morning. Joined by three natives, Humboldt and Bonpland headed for the gigantic pyramid. The slope was gradual at first, and their progress was quick, but after they encountered snow at 4,400 meters (14,440 feet), the trail became more treacherous. At a little over 5,000 meters (16,400 feet), the path narrowed and steepened. From that point on, the slopes were icy and steep, and the men needed to hold on to rocks to keep their balance. Their hands bled from grabbing the sharp rocks, and their shoes were soaked with melted snow. Even worse were the effects of high altitude: nausea, vertigo, difficulty breathing, and bleeding from the nose and the gums. Yet none of this perturbed them. Humboldt and Bonpland knew the symptoms of altitude sickness well, and they knew how far they could go.

Since early that morning, the expedition had been proceeding through dense clouds and mist, preventing the men from getting a view of their surroundings. They were relieved when, at noon, the clouds opened up and the immense, dome-like shape of Chimborazo's summit appeared just before them. It was "a somber, magnificent view," as Humboldt recorded in his diary.[1] He quickly grabbed the cyanometer from his bag to measure the sky's blueness.

Literally a "measurer of blue," the cyanometer (Color Plate 15) was a scale of fifty-two shades of blue ranging from white ("zero degrees") to black ("51 degrees") via an ultramarine hue ("Berlin blue"). This simple device had been invented in the 1760s by Horace-Bénédict de Saussure, a Swiss geologist and Alpine explorer. Saussure distributed the scale, drawn with water colors following a well-defined recipe, to friends and fellow naturalists with the aim of compiling a comprehensive record of the sky's blueness at different locations, elevations, and times. Observers were to proceed in a set manner, holding the scale up to the sky with the sun at their backs. Soon they discovered that a normal, midday, clear blue sky at sea level typically measured 23 degrees on the cyanometer's scale. This was a darker hue, and thus a larger number, than in the morning or afternoon. It was known that the sky appears even darker when viewed from a higher altitude. This phenomenon may have been discovered in the tenth century by the Arabian explorer Ibn al-Biruni, who had attempted to climb Mount Demavend, the highest mountain in Persia. Saussure was able to quantify this finding on a clear day in 1787, when he took the cyanometer along to the summit of Mont Blanc, Europe's highest mountain. There he recorded a blue of 39 degrees. Eight years later, while preparing for his expedition, Humboldt visited Saussure in Geneva to inquire about the cyanometer and to obtain a well-calibrated copy. On his expedition, Humboldt became an avid user of the scale. He used it not only on the voyage across the Atlantic Ocean, where he recorded "23.5 degrees at noon overhead" in the Caribbean, but also at the summit of Teide in the Canary Islands, where he noted a record 41 degrees on the scale.[2]

When the cloud cover over Chimborazo broke during that brief interval at noon on June 23, 1802, at an altitude never before reached by humans, Humboldt set a new record for the darkest blue sky ever seen: 46 degrees on the cyanometer. After a few minutes the cloud cover returned; the members of the expedition were again covered in mist, and neither the summit nor the plains below were discernible. Nevertheless, they continued their ascent. But shortly afterward, they hit a dead end: at an elevation of 5,878 meters (19,285 feet) an "insurmountable cleft" blocked their path. Four hundred meters below the summit, they had no choice but to turn back. Humboldt and Bonpland were disappointed but happy. They had failed to reach the summit, but they knew that nobody had ever before climbed to such a high altitude. While they retraced their steps down the mountain, weather conditions worsened. It began to hail, and then to snow. At five o'clock in the afternoon they reached Calpi, exhausted, and stayed another night with the village's friendly priest. The next day, Humboldt dryly noted in his diary, "As usual, the cloud-covered day of the ascent is followed by the fairest clear skies."[3]

Humboldt's obsession with taking measurements of all the phenomena that could possibly be expressed in numbers is best understood in the context of Enlightenment natural philosophy. In that era, expeditions set out to every part of the world, from Saussure's journeys in the Alps to James Cook's sailing cruises in the South Seas. These voyages produced an unprecedented amount of data, which was left to the natural philosophers back home to make sense of. Scholars were eager to compare the various findings. Taking measurements that could be expressed in numbers seemed the most promising way to do so.

It is worth noting that by 1814, a decade after returning from his expedition, Humboldt had grown critical of the cyanometer's usefulness, calling it an "instrument that is still incomplete."[4] Eventually it became clear that the scale would not help to unravel the mystery of the sky's blueness. Saussure had been certain that it would; he believed that the blue color was due to the hue of the air's moist particles, a hue that he estimated at 34 degrees on his cyanometer's scale. If this were so, the instrument would provide an estimate of how many of these particles were floating in the air over a given place. But aside from the awkwardness of this hypothesis, the cyanometer itself proved inadequate to fulfill its purpose: comparability. Observers began to suspect that the different levels of illumination provided by the sun, inevitably present at every measurement, confounded the instrument's readings. And in another way, the cyanometer failed because the color of the sky is greatly influenced by particles that are neither blue nor moist.

Humboldt's fame as a mountaineer was to last longer than interest in the cyanometer. After that historic day in June 1802, almost three decades passed before anyone else attempted to climb Chimborazo. On December 16, 1831, the Englishman Obrist Hall managed to reach a point only 64 feet higher than Humboldt and Bonpland. At last, in 1886, the Italian brothers Jean-Antonine and Louis Carrel and the Englishman Edward Whymper reached the mountain's summit. With an altitude of 6,310 meters (20,700 feet), Chimborazo is indisputably the highest mountain in Ecuador. However, by the time Hall, the Carrels, and Whymper climbed it, it had lost its status as the world's highest peak. During the period from 1819 to 1825, the Trigonometric Survey of India had established that some of the mountains in the Himalayas were higher than any peak in the Andes, with Dhaulagiri, in today's Nepal, reaching an altitude of 8,558 meters (28,077 feet), and Kanchenjunga, on the border between India and Sikkim, allegedly 8,588 meters (28,176 feet) high. The mountain that would later be called Mt. Everest was not discovered by British surveyors until 1856.

MYSTERIOUS BLUE SHADOWS

If measuring the sky's blueness was one obsession of Enlightenment naturalists, debating reports about perplexing natural phenomena was another. Otto von Guericke, the mayor of Magdeburg and inventor of the air pump, describes an astonishing observation in his 1672 book, *New Experiments from Magdeburg About Empty Space* (Neue Magdeburger Versuche über den leeren Raum). One morning at sunrise, Guericke saw the shadow of his finger cast by the light of a candle onto a white page. The shadow did not look black or gray, as he had expected, but blue. Almost two centuries earlier, Leonardo da Vinci had noticed shadows of a similar color, but not until the middle of the eighteenth century were they systematically observed and made the topic of scientific discussion. The pure blue color of the shadows seemed to defy common sense, sparking a debate about what could cause it.

Colored shadows seem paradoxical because most people believe that all shadows occurring in nature are gray or black, thus lacking color. However, astute observers can often see colored shadows. The French naturalist Georges-Louis de Buffon wrote that anyone can observe this phenomenon if he casts the shadow of a finger onto a white sheet of paper when the sun is low in the sky. Toward evening, the shadows seem to be tinged green rather than blue. The German author Johann Wolfgang von Goethe noticed blue shadows on December 10, 1777, when he came down from the Brocken, a peak in the Harz mountain range in northern Germany. Under a clear sky, Goethe noticed that all the shadows around him looked blue on the white snow.

A few years later, the Swiss naturalist Nicolas de Beguelin gave a lecture before the Berlin Academy of Sciences in which he presented an explanation of this phenomenon (Figure 5.2). He noted that the observations of Guericke, Buffon, and Goethe agreed in that whenever the sun is low and the sky is clear, a blue shadow appears if the object that is casting the shadow is illuminated by reddish-yellow sunlight. Beguelin knew that when the sun is just above the horizon, the clear sky is still lit up blue at the zenith, and it illuminates the paper, even in the area where

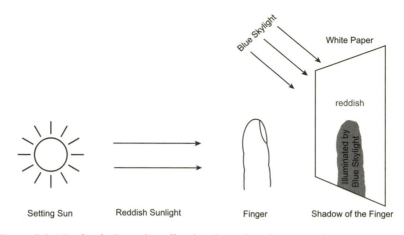

Figure 5.2 Nicolas de Beguelin offered a physical explanation of the blue shadows in 1767. The setting sun casts the shadow of a finger onto a sheet of white paper. The clear blue sky illuminates the shadow and gives it its color.

the shadow cast by the sunlight is located. And so the light from the sky colors the shadow blue, whereas the reddish sunlight has no chance to impose its color on the shadow. On the other hand, in the part of the paper that is lit up by the sun, the light from the sky is outshone by that of the sun, so that only the color of the direct sunlight is visible there. Beguelin concluded that this is why the shadow looks blue, while the paper looks reddish yellow. Today, you can see how plausible this explanation is when you attend the theater. When the stage is lit up from both sides by different-colored lights, we see the shadows cast by the light of one lamp in the direct light of the other, and so they appear in the color of the latter.

Most contemporary researchers were satisfied with this explanation of the blue shadows. Goethe (Figure 5.3), however, did not share Beguelin's view. Thirteen years after his trip into the Harz Mountains, he conducted an experiment showing that blue shadows could be seen even without illumination by the blue light of the sky (Figure 5.4). The experiment calls for covering the window of a room with a white curtain. To make sure that the clear sky's blue light does not influence the observations to follow, Goethe recommended performing the experi-

Figure 5.3 Portrait of Johann Wolfgang von Goethe by Joseph Karl Stieler (1828). Courtesy of Bildarchiv Preussischer Kulturbesitz, Berlin.

ment on an overcast day with no trace of blue in the sky. To prevent any further influence of blue color perceptions, he also strongly recommended removing anything blue from the room. Having done this, one should light a candle, use it to illuminate a sheet of white paper held

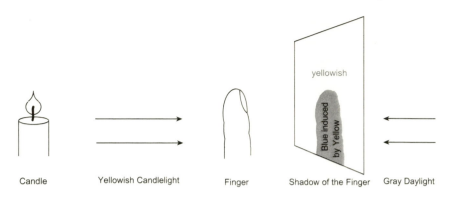

Figure 5.4 Goethe disagreed with Beguelin's explanation of the blue shadows and suggested a physiological explanation instead. The yellow light of a candle illuminates the white paper, which is lit up from the other side by gray daylight. Although the shadow of the finger is not lit up by the candle, nonetheless the blue color (being complementary to yellow) is "induced" in the eye.

against the window, and observe the shadows cast on the sheet by an intervening object. These shadows, Goethe claimed, will be perceived as tinted blue.

Having systematically eliminated any direct influence of blue skylight, Goethe was thus able to show that there must be alternatives to Beguelin's interpretation. Four years later, the English physicist Benjamin Thompson stumbled onto a similar finding. Experimenting with the shadows of an object cast by two candles on either side of it, Thompson was able to eliminate the influence of daylight entirely. While Thompson saw the colored shadows as a mere curiosity, this baffling experiment prompted Goethe to undertake a study of the nature of color, for if the blue color of the sky is not involved in creating the blue shadows, what else could cause them?

Where Beguelin had proposed a physical cause, Goethe sought a physiological explanation for the colored shadows. In other words, he deals with the actual impressions we perceive. We have seen that blue and yellow are opposite each other on Newton's color circle, so that they form a pair of complementary colors. Likewise, Goethe named blue/violet and yellow/red as opposing color poles. The reddish-yellow candlelight makes the white paper appear as a reddish-yellow surface, says Goethe. In contrast, the shadow is unaffected by the color of the candlelight, but the neighboring reddish-yellow surface "induces" the eye to see it as blue—the color that is the polar opposite of yellow/red. What Goethe described as the "induction of the color blue" is today known as simultaneous contrast. If you look at a white surface next to an intensely colored surface, the white surface appears to take on the complementary color of the colored surface. Goethe was right in criticizing Beguelin's explanation as incomplete, because simultaneous contrast does play as important a role in our perception of the blue shadows as the physical causes cited by Beguelin. Consistent with Goethe's explanation, replacing the candles with a blue light source makes the shadows appear yellow.

Neither Goethe nor Beguelin could know that other physiological processes also affect how we perceive the colored shadows. If you stare intently at a colored surface for a while and then look at a white surface, the latter will appear in the complementary color of the colored surface. This is called an after-image. And finally, a phenomenon known to psychologists as color constancy influences our interpretation of shadows. Because we usually expect shadows to be gray or black, we perceive them as gray or black. The brain plays the trick on us of correcting color impressions from the eyes that it assumes to be false because they do not match its expectations. While simultaneous contrast and after-images amplify the perceived colors, color constancy weakens them. It thus helps us to create order out of the chaos of visual impressions that we are exposed to every day. After all, the kind of light that illuminates the objects around us is changing constantly. It is thanks to color constancy that we can recognize a white surface as white, even under the yellowish light of a light bulb or a candle.

AN OPTICAL ILLUSION?

The puzzle of the blue shadows gave rise to a new way of looking at the blue color of the sky. Benjamin Thompson had considered the color of the blue shadows to be an optical illusion, and wondered whether the same might be true of the sky's blue. In the 1820s Georg Wilhelm Muncke, a German physicist from Heidelberg, thought that he could confirm this conjecture. He wrote in an essay:

> Most interesting is the well-founded conclusion from this line
> of thinking, that the atmospheric air is not blue-tinted, as has been
> generally assumed, thus producing the blue color of the sky across
> the greater thickness of the atmosphere; rather, this color is merely
> subjective and complementary to the very white sunlight, tend-
> ing toward yellow, which is reflected off the earth and the objects
> on it.[5]

Muncke's experiment and his conclusions caused something of a sensation. This was surely due to the fact that anyone could easily repeat the experiment. He took a long pipe, blackened on the inside, and looked through it with one eye at the blue sky. With the other eye he looked directly at the sky. After a while, the sky's blue as seen through the pipe began to appear lighter and lighter, eventually fading to white. Meanwhile, the naked eye continued to see the blue unchanged. Muncke concluded that the pipe had shut out the influence of outside light so that the view through it revealed the sky in its true white color. The naked eye, said Muncke, is tricked by the light of the sun and the landscape and conveys the impression that the sky is blue, apparently tinged with the complementary color of yellow sunlight.

Though Muncke intended to expose the blue color of the sky as an illusion, he instead fell victim to an optical illusion himself. A few years later Heinrich Wilhelm Brandes, an astronomer and physicist from Breslau, Germany, showed that Muncke's experiment could not stand up to careful scrutiny. Brandes lived in an apartment whose light-blue wallpaper approximated the blue hue of the sky. He repeated Muncke's experiment, substituting the wallpaper for the sky. After he had looked at the wallpaper through the pipe for a while, it too appeared to lose its color and produced more and more of a white impression. Looking at the blue sky through the pipe led to the same result, and so Brandes concluded that he would be as justified in calling the sky blue as he would his wallpaper, whose color was beyond question. Johann Wolfgang von Goethe also repeated Muncke's experiment, reaching the same conclusion as Brandes:

> June 23, 1822. Repeated the Muncke experiment under a deep blue sky and found what will and must always be found: the blue color unchanged in the naked eye, lighter in the shielded one; but the same goes for the clouds, the spruce woods, the whole surrounding area.[6]

Brandes and Goethe recognized that Muncke's experiment did not constitute grounds for doubting the reality of the sky's blue color. Goethe's reaction to Muncke's claim was especially fervid. He found the latter's statements cause "for sad reflections," because they would "retard the dissemination of the true color theory."[7] Perhaps inadvertently, the Heidelberg professor had attacked the foundation of Goethe's *Theory of Colors*, which had appeared in print ten years earlier. Goethe regarded the blue of the sky as an "ur-phenomenon," or basic phenomenon, in which an explanation of all the colors of nature was concealed. Anyone who doubted the reality of that blue color called the entire theory into question—and made himself the target of Goethe's polemics.

BASIC PHENOMENON

> The ultimate insight is that what is factual is already theory. The blue color of the sky reveals to us the fundamental principle of chromatics. Therefore don't look for anything behind the phenomena— they themselves are the lesson.[8]

This is what the seventy-nine-year-old Johann Wolfgang von Goethe wrote in his diary in 1828, four years before his death. Today we know Goethe more for his literary works, which include *Faust* and *The Sorrows of Young Werther*. But the German writer also did research in natural science. This can be seen in his investigation of the blue shadows, which ushered in Goethe's systematic study of color theory. His interest in this subject probably dates back to his college years in Leipzig. There he had

begun studying law as a sixteen-year-old in 1765. On the side, he nur-
tured an interest in natural science, attending Johann Heinrich Winckler's lecture "Foundations of Physics." Winckler was critical of Newton's
theory, and although he dealt with optical phenomena in detail, he
made only glancing references to Newton's important discoveries. In a
letter dated 1769, Goethe mentions that thinking about light and color
was one of his favorite pastimes. From September 1786 to the middle of
1788 he took an extended trip to Italy, which brought him to Venice,
Rome, and Naples, among others. In his diaries his interest in the blue
color of the sky shines through. For example, on October 8, 1786, he
describes the "blue air" over the Venice Lagoon enthusiastically.[9] In the
entry from February 9, 1788, we find the comment that he had read
"Leonardo da Vinci's book about painting,"[10] He was referring to the
Treatise on Painting, which had acquainted the world with Leonardo's
studies on aerial perspective. This whetted Goethe's curiosity. Soon
after returning from Italy, he began a systematic study that would culmi-
nate in his 1810 book, *Theory of Colors*. Right from the beginning, he
focused his studies on the "physiological" colors that we perceive, such
as those of the blue shadows. Almost instantly he came to oppose Newton's optics:

> At the same time, I turned my attention to artists' color tech-
> niques, and when I went back to the basic physical elements of
> this theory, I discovered to my great amazement: the Newtonian
> hypothesis was wrong and untenable.[11]

How dare this poet criticize the optical theory of a world-famous physicist? The impetus for Goethe's critique was as trivial as it was far-
reaching. As Goethe explained, after returning from his trip to Italy, he
began to focus on how artists use color. Alongside this he endeavored to
understand the fundamental physical principles of colors, which inex-
orably led him to Newton's prism experiments. Goethe borrowed several
glass prisms from his friend, court counselor Büttner of Jena, in order to
repeat these well-known experiments. A few months later, Büttner had
not gotten his prisms back, and so he sent a messenger to Weimar to re-
trieve them. Goethe had long since forgotten the prisms and was about

to return them when he caught a fleeting glimpse through a prism at the white wall next to his threshold—and was astounded to find that he could not see the colors of the spectrum that Newton had written about. On the contrary, the wall looked just as white through the prism as it did to the naked eye. The messenger had to leave empty-handed, for Goethe did not want to let the prisms go without further investigating this "discovery." To him, it was proof that white light was not a mixture of the colors of the spectrum, as Newton had claimed, but an indivisible basic color, as Aristotle had written.

For Goethe, the fleeting glimpse through the prism sufficed to call Newton's optics into question. But Newton had never claimed that one could see the colors of the spectrum by holding a prism up to one's eyes. Newton had only investigated the colors of the spectrum under the special conditions of an experiment in which the prism projected the refracted light rays onto a wall several yards away. Within Newton's framework, Goethe's observation was no surprise; seen from up close, the refracted, different-colored light rays would overlap each other so much that their mixed light would show the colors combined. If you hold a prism *directly* in front of your eyes and look at a white wall, then even according to Newton it must look white.

Goethe, however, thought he could show that the emperor had no clothes. He performed a series of additional tests, in none of which Newton's experiments were replicated exactly. Rather, Goethe always held the prism right up to his eyes. As a proponent of Aristotelian philosophy, he saw his experiments as a confirmation of the Greek's 2,000-year-old color theory. Goethe thought he could prove that theory with his prism experiments, though as far as most physicists were concerned, Newton's prism experiments had refuted it once and for all. Goethe saw the colors of the spectrum most clearly when he looked through the prism at the border between neighboring black and white surfaces. Wasn't this proof that all colors result from the combination of black and white, and that Aristotle had been right?

Without a doubt, this was a case of two fundamentally different views of nature colliding. For Newton, the prism was a scientific device that split sunlight into rays of varying refrangibility and made it possible to analyze the physical properties of light. His prism experiment used an indoor room to isolate the sunlight from the outdoor world, in order to reveal its secrets. Goethe, on the other hand, saw the prism less as a scientific device and more as a part of nature with the capacity to clearly illustrate one of nature's general characteristics. You don't need any instruments to understand the essence of nature, he writes, because the human being himself is the best instrument for the task:

> There can be no greater or more exact apparatus for physics than the human being himself, provided he makes use of his healthy senses. And this is precisely where modern physics has its greatest failing: that the human being has been quasi eliminated from the experiments, so that nature is only observed by taking readings with artificial instruments; indeed, people have sought to constrain and prove what nature is capable of through these means.[12]

In Goethe's worldview, the human being can directly apprehend nature itself, whereas "modern physics" cuts nature up into pieces, analyzes them and conducts laboratory experiments that can only be replicated under very special conditions. The price for these insights gained in a laboratory was far too high, in Goethe's view; after all, he saw them as banishing everything human from natural science and ultimately destroying nature itself. This is the only way to explain his often repeated polemics against Newton and his devotees. Along the same lines, he gave the second part of his *Theory of Colors* the title *Newton's Theory Exposed*.[13] Like a Don Quixote fighting windmills, the German poet was standing up against an experimental tradition of physics that had begun with Galileo Galilei. His response was much different from that of contemporary English poets, who considered the *Opticks* Newton's most comprehensible and intuitive work. For them, it was a source of inspiration—metaphors and ideas on light and color soon abounded in poetry, from which they had been missing since the early seventeenth century.[14]

Goethe criticizes Newton's physics vigorously, but what alternative does he have to offer? To him, the best example of the human ability he touts—that of directly perceiving nature—is seeing the blue color of the

TABLE 5.1

EXAMPLES OF THE OPTICAL EFFECT OF TURBID MEDIA FROM THE
DIDACTIC PART OF GOETHE'S *Theory of Color* (1810)

Turbid Medium	Background	Color
Cataract (in the eye)	Illuminated objects	Red
A certain degree of haze	Sun	Yellow
A larger mass of haze	Sun	Red
Haze lit up by daylight	Darkness of infinite space	Blue
A thickening haze	Darkness of infinite space	Whitish blue
Air with fine haze	Distant mountains	Blue
Lower part of a candle flame	White	White
Lower part of a candle flame	Black	Blue
Liqueur	Dark wooden cup	Blue
Liqueur in a glass	Sun	Yellow
Parchment pages	Sun	Yellow to red

SOURCE: Johann Wolfgang von Goethe, "Zur Farbenlehre," vol. 10 in *Sämtliche Werke nach Epochen seines Schaffens, Münchner Ausgabe* (Munich: Hanser Verlag, 1989).

sky. It plays a central role in his color theory, illustrating as it does the basic phenomenon of the optical effects of turbid media. Goethe would explain the colors of the spectrum on this basis. Although he had characterized the blue shadows as a physiological color, to him the basic phenomenon belongs to the realm of physical colors.

Turbid media, according to Goethe, comprise all materials and media that are partly transparent. In nature, they are more the rule than the exception. We encounter them every day: soapy water, liqueur, frosted windows, the bottom half of a candle flame, parchment paper, Earth's atmosphere—these are examples used by Goethe in the *Theory of Colors* to demonstrate the diversity of such media. Though these materials differ in various ways, they have the same optical effect (Table 5.1). Goethe distinguishes between two cases: *bright* objects, when seen through a turbid medium, look yellow or red. On the other hand, when a *brightened* turbid medium is seen in front of a dark background, it seems to be blue (Figure 5.5). The degree of turbidity determines the color that results.

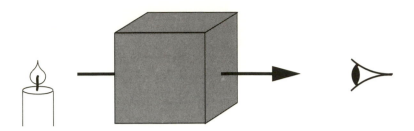

Bright objects look yellowish-red if viewed through a dark turbid medium

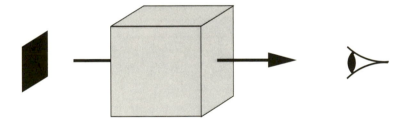

Dark objects look bluish-violet if viewed through a brightened turbid medium

Figure 5.5 Goethe's basic phenomenon describes the two ways in which turbid media can produce the colors blue/violet and yellow/red.

Media that are only slightly turbid barely affect the apparent color of bright objects behind them. With increasing turbidity, however, they make them appear yellow or reddish. Slightly turbid media that are lit up make distant, dark objects appear blue. Yet with increasing turbidity, this color becomes progressively paler. Goethe calls this effect the basic phenomenon (*Urphänomen*) of chromatics, which is best visible in the clear daytime sky:

When we view the darkness of infinite space through atmospheric hazes illuminated by daylight, the color appears blue.
From atop high mountains, the sky looks royal blue during the
day because there is only a little bit of fine haze floating in front
of the infinite dark space; as you descend into the valleys, the blue
becomes lighter until finally it makes the full transition in certain
regions, and with increasing haze, to a bluish white.
Mountains, likewise, look blue to us; for when we view them from
far enough away that we can no longer see the local colors and no
light from their surface affects our eyes, then in effect they become
nothing more than dark objects that only appear blue due to the
intervening hazes.[15]

In the sky, Goethe contends, the "atmospheric hazes" act as a turbid
medium, because they are illuminated by the sun in front of the darkness
of outer space. That is why we see them as having a blue color. At sunset,
the second optical property of turbid media reveals itself: the sun, as it
shines low in the sky, appears yellow or even red through the atmospheric hazes. Both properties of the basic phenomenon are thus mani-
fested in the sky, and we do not need any tools to see the "theory of colors" in action.

In mentioning the apparent coloring of distant mountains and the
darkness of outer space, Goethe's work is reminiscent of Leonardo's *Treatise on Painting*. The above quotation reads as if Goethe had been copying
straight from the master painter's notebooks. Yet in fact, Goethe takes it
beyond where his predecessor left off and generalizes Leonardo's observations. He notices that the blue and yellow/red colors that are visible in
turbid media correspond exactly to the two color poles he had identified with regard to the colored shadows. Yellow and blue, as the positive
and negative poles, respectively, are the parents of all colors. But how
do the blue and yellow poles produce the other colors—violet, green,
orange, and red? To explain this, Goethe returned to his prism experiments. He put forward an explanation of the color spectrum that was
intended to replace Newton's interpretation, and which he tried to derive from the basic phenomenon (Color Plate 16). You take a glass prism

and look at a rectangular white area against a black background. Goethe says the prism seems to shift the black area, blurring it at the same time. One edge of the black area ends up superimposed over the white background, and so it appears blue. The opposite edge is darkened by the turbid medium of the prism and looks yellow, while the middle remains black. On the edge next to the blue, violet appears. Next to the yellow we find orange. If the black area is quite narrow, the colored edges can overlap so that yellow and blue combine to make green. Red results from the "augmentation" of violet and orange. By using the concepts of lightness, darkness, blurring, and augmentation, Goethe has thus explained the colors of the visible spectrum from violet to orange. Consequently, for him there are six colors: the two elementary colors yellow and blue, the mixed color green, the peripheral colors violet and orange, as well as the augmented or amplified red.

Goethe's first draft of the *Theory of Colors* was completed around 1800. It already contained his views on the blue color of the sky as due to the air being a turbid medium. Yet it was not until ten years later that he published the *Theory*. When he finally delivered the last pages to the printer, on May 16, 1810, he called it a "joyous day of relief."

The book divided his readership into two camps. On one side were its devoted admirers, like the English painter J. M. W. Turner, who carefully annotated his copy of the book. Turner found Goethe's ideas useful for the artist's practice and celebrated them with the painting *Light and Color (Goethe's Theory)*. Contemporary experts were highly impressed with Goethe's extensive and original observations on the physiological colors. On the other hand, his interpretation of the spectral colors was received less favorably. It earned him the criticism and derision of contemporary physicists, who accused him of meddling in a field he knew nothing about. But he held firm to his position, seeing himself as the lone defender of a general truth. The significance of the theory of colors, he insisted, went far beyond the study of chromatics and was a model for future, holistic studies of nature.

Later scientists have evaluated the *Theory* much more positively than Goethe's contemporaries did, and they did so for a variety of reasons. Take Hermann von Helmholtz, for example, one of the most famous physicists of the nineteenth century. Though Helmholtz was forced to admit that Goethe had failed to produce any valid physical arguments against Newton's optics, he nonetheless praised the expressiveness with which Goethe had described the physiological colors. This was an endeavor that Helmholtz himself was actively involved in. Werner Heisenberg, one of the founders of quantum physics, talked about Goethe's *Theory* in an influential lecture.[16] He came to the conclusion that Newton's and Goethe's theories simply dealt with different things, and so it made no sense to pit them against each other. Both approaches deserved to be pursued further, said Heisenberg, who sensed that Goethe had anticipated important aspects of a modern view of nature. The principle of symmetry, which is exemplified in the blue and yellow poles of the *Theory of Colors*, resembles concepts from modern particle physics. But where Goethe criticized Newton, it was beyond even Heisenberg to defend him.

From their comparative study of the experimental methods employed by Goethe and Newton, historians of science Neil Reid and Friedrich Steinle conclude that Goethe's method in *Theory of Colors* is a representative example of what they call "exploratory experimentation." Looking through the prism, Goethe was initially faced with a view of enormous complexity. The prism had distorted the colors of the landscape into a bewildering array, and it was not immediately clear to him how that came about. To make a long story short, Goethe proceeded by reducing the complexity of the observational situation, introducing a series of hypotheses along the way. One idea was that the colored fringes appear at the border between dark and bright regions. To investigate this hypothesis he drew a dark rectangle on a white sheet of paper, and then looked at it through a prism held to his eye. As I noted earlier, this helped him perceive the regularity of the colored fringes and formulate rules predicting their occurrence. Proceeding in such a way to derive simple situations from a complex scenario, Goethe's "exploratory experimentation" does not presuppose a theory of light and color. Reid and Steinle point out that a similar approach was taken by Michael Faraday and André-Marie Ampère, the two most famous experimenters on electricity in the nineteenth century. In contrast, Newton's optical experimentation was theory-oriented in that he assumed light to propagate as rays along straight lines, those rays being of different refrangibility. This initial assumption was guiding Newton throughout his series of experiments and features again in his writing of *Opticks*.

Eventually, Goethe was forced to acknowledge some serious problems with his color theory. For one thing, doubt was cast on the universality of the basic phenomenon. Heinrich Wilhelm Brandes, who had already debunked the idea of the sky's blue as an optical illusion, tried in 1827 to demonstrate the colors of the basic phenomenon using water vapor and air saturated with mist. But although both of them are turbid media according to Goethe's definition, the expected color phenomena were nowhere to be seen. Moreover, the proofs of Newton's theory could not be dismissed by Goethe's simple experiments. While Newton could easily predict the sequence of colors in the rainbow using his theory, Goethe never managed it, in spite of numerous attempts. He finally agreed to leave out the "second, polemical part" of the *Theory of Colors*, which he had titled *Newton's Theory Exposed* in 1810.

THE UNIFORMITY OF AIR

After traveling for five years, Humboldt and Bonpland returned to France. The two explorers were received with admiration and curiosity. Throughout their absence, academic circles in Europe had eagerly awaited any news of their exploits. After their adventure on Chimborazo, the two had moved south to Peru and then north to Mexico, Cuba, and the United States, before arriving in Bordeaux on August 21, 1804. Humboldt and Bonpland were on their way to Paris when, on August 24, the physicists Louis Joseph Gay-Lussac and Jean Baptiste Biot made a memorable ascent in a hot-air balloon, rising to a record altitude of 4,000 meters (13,000 feet). Only three weeks later, on September 16, they made yet another ascent, reaching the staggering height of 7,000 meters (23,000 feet). This time, Gay-Lussac and Biot took samples of the air in bottles, which they sealed immediately. These samples were intended to help answer an intriguing question: does the chemical composition of air vary from place to place, or is it the same throughout the atmosphere?

This question was well posed. Back in the 1660s, Robert Hooke had surmised that the air consisted of at least two parts, associating one of them with the processes of combustion and animal respiration. Yet only in the 1770s was this part, now called oxygen, investigated in greater detail by the chemists Henry Cavendish, Carl Wilhelm Scheele, and Antoine de Lavoisier. In their laboratories, they found that it amounted to about 27 or 28 percent of the air's volume, the remaining percentage consisting of chemically inert nitrogen. But were these percentages the same everywhere and at all times? Perhaps not, Humboldt had suggested eight years before; he believed that the percentage of oxygen decreases in the atmosphere with increasing altitude.

In late 1804, Gay-Lussac joined forces with Humboldt to investigate this question. With remarkable ease, Humboldt switched from being an expedition naturalist to being a laboratory scientist. With a refined technique, the two found that Cavendish, Scheele, and Lavoisier had arrived at incorrect proportions for the constituents. The correct values appeared to be 21 percent oxygen gas, 78.7 percent nitrogen gas, and 0.3 percent carbonic acid, a trace constituent that had been overlooked in previous investigations. By and large, these fractions are the accepted values today. The samples taken during the balloon ascent revealed that this composition does not seem to change with altitude. Humboldt and Gay-Lussac, who soon became close friends, investigated this issue further by taking additional samples of the air in Paris, with the wind coming from different directions. In each measurement, the percentages were the same. Eventually, they visited Mont-Cenis, a monastery in the Alps at an altitude of 2,650 meters (8,700 feet), taking along with them a bottle of Parisian air, which they compared with the Alpine sample in their mobile laboratory. Once more, the percentage of oxygen was the same as in each of the previous measurements; the two scientists concluded that the composition of air is constant throughout the lower atmosphere. Two decades later, this finding continued to amaze, as seen in Johann Samuel Traugott Gehler's *Physical Dictionary* of 1825:

The main constituent parts of atmospheric air are nitrogen
gas and oxygen gas, which—according to unquestionable
experiments—are combined in a very constant quantitative proportion in all regions of the Earth, in all seasons, in high and low
elevations, in the outdoors as well as inside rooms and even in
opera houses and hospital wards.[17]

Constant as the chemical composition of the air may be throughout the
lower atmosphere, there remained the possibility that it may change at
intervals longer than the course of the seasons.

CHAPTER 6

A Polarized Sky

In August 1851, ten live African chameleons arrived at the Imperial Academy of Sciences in Vienna, Austria. The animals had been collected in the desert near Cairo, Egypt, by Dr. Lautner, a corresponding member of the academy. A few months earlier, the Academicians had pondered the color changes that occur in chameleons' skin, concluding that this phenomenon remained a riddle, but one worth solving (Figure 6.1). Various answers had been proposed since antiquity. Aristotle surmised that the color change occurs only when these reptiles inflate their throats or when they die. The Roman poet Ovid was among the first to claim that chameleons assimilate to the color of their surroundings, still a popular belief today. Others suspected that the color changes occur when the chameleons become angry or are frightened, when they are exposed to sunlight, or that these normally black animals change color when suffering from yellow fever. Clever experiments had been devised. For instance, in 1829, the Swiss Robert Spittal held a candle close to a sleeping chameleon and noticed that large brown spots appeared on its skin, even while it remained asleep. Upon removal of the candle the spots disappeared. Spraying chameleons that were awake with water also yielded such spots. Spittal concluded that the animals' color could change voluntarily as well as involuntarily.

The chameleons' skin was the place where the color change occurred, and so it seemed that a microscopic investigation could yield new insights. This at least was the academicians' hope when they placed their order with Dr. Lautner, who was known as a reliable procurer of flies, spiders, and parasites from the Egyptian desert.

Once the ten chameleons (of the species *Chamaeleontis viridis*) had arrived in Vienna, four of them were given to the Imperial Court's Natural History Cabinet. The other six were submitted for examination to

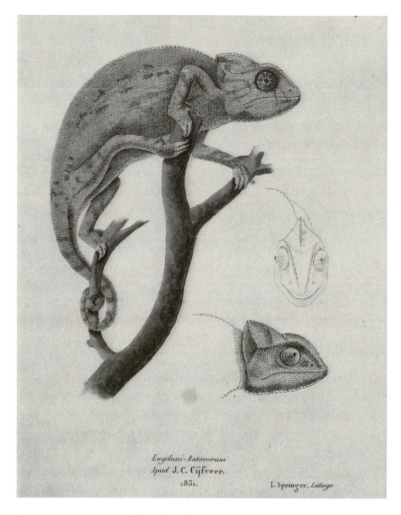

Figure 6.1 Sketches of the chameleon's color changes by Johann van der Hoeven. Reproduced from van der Hoeven's *Icones ad illustrandas coloris mutations in Chameleonte* (Leiden, 1831), Plates I and V.

the laboratory of Ernst Wilhelm von Brücke, a renowned professor of physiology who had moved to Vienna two years before. The son of a painter, Brücke (Figure 6.2) had studied medicine and physiology in Berlin and became a professor at the University of Königsberg in 1848.

Figure 6.1 (*continued*)

His renown as a scientist stemmed from his skillful application of the microscope in studying physiological processes, and thus he was the man to investigate the chameleons.

Brücke began by carefully observing the living animals' skin under his microscope. Upon wetting the skin with his saliva, he noticed a colorful shimmer that seemed to originate in the epidermis, the uppermost part of the skin. He then proceeded to kill an animal and remove the

Figure 6.2 Portrait of Ernst Wilhelm von Brücke from 1890. Courtesy of Hans von Brücke.

epidermis layer by layer. Part of the skin's apparent coloration reminded him of the sequence of colors recorded by Newton when pressing two glass lenses together. Brücke noticed a mix of colors that was dominated by the blue and yellow hues of Newton's second order. Nevertheless, the epidermis was partly transparent, and so Brücke concluded that its color may combine with that of deeper layers. Having removed the chameleon's epidermis, Brücke reached the cutis. Looking through his microscope, he discovered that this subsurface skin layer contained two layers of pigment cells. The cells in the upper layer were filled with a light-colored (white or yellow) pigment, while the layer below this consisted of inky black pigment cells. Brücke determined that these black cells may recede so far down beneath the surface that only the light pigments remain visible. In this case the chameleon's skin appears white or yellow. In response to nerve impulses, the black pigment cells can move closer to the surface. They then show through the light pigment cells, making the skin appear blue-gray to violet. If yellow pigment cells are layered over this, then the skin appears green; on the other hand, if the black pigment cells in the cutis migrate almost all the way up to its surface, then the chameleon looks black. The shimmering colors of its epidermis play only a subordinate role in these color changes.

In unveiling the mechanism behind the chameleons' color changes, Brücke noticed its similarity to Goethe's explanation of the color of the sky: both are the optical effect of turbid media. Spurred on by his studies with the chameleon, Brücke began to explore these media more generally. In doing so, he was strongly influenced by Goethe's *Theory of Colors*, a book that Brücke had read with enthusiasm in his student days. Like Goethe before him, he knew that turbid media are commonly found in nature: the atmosphere, seawater, and the skin of the chameleon are just three examples. Soon the German scientist Hermann von Helmholtz, a friend of Brücke, realized that even the colors of human eyes are an effect of turbid media (see Appendix B).

The atmosphere could be considered a turbid medium if small particles are suspended in it. One effect of these particles is to disperse the light of the sun throughout the sky, thus producing diffuse daylight. Where Goethe considered all "not completely opaque materials" as turbid media, Brücke strived for a stricter definition. He defined them as a combination of two or more media of differing indices of refraction whose smallest particles are not visible to the naked eye, and which only reveal themselves through the light that they reflect and disperse. Where Goethe had regarded the optical effect of turbid media as a basic phenomenon whose causes could not be further dissected, Brücke attrib-

uted it to the reflection and scattering of light and thus sought to identify a physical mechanism behind the basic phenomenon.

Brücke was able to confirm Leonardo's observation that smoke looks
blue in color when seen in front of a dark background while lit from the
side, but that it appears reddish or ashen when the light source is directly
behind it. Brücke had observed the exact same color phenomena in a
number of turbid media, and so he proceeded to study them under the
controlled conditions of his laboratory. The suspension of mastic, the
resin of a Mediterranean pistachio tree, in ethanol was to be his breakthrough experiment (Color Plate 17).

> Take a solution of one gram of the finest, most colorless mastic in
> 87 grams of ethanol, and drip it into water kept in constant mo-
> tion through vigorous stirring. This yields a cloudy liquid which
> shows the . . . color phenomena perfectly. The blue inks can be
> seen best when poured onto a tablet of black glass or into a black
> bowl; in order to study the yellow and red ones, pour them into a
> bottle with parallel sides, then look through it at brightly lit white
> objects, at the sun itself, or at a flame. In this liquid, . . . the reflec-
> tion and scattering of light have the very same proportions as they
> do in the atmosphere.[1]

Recognizing this observation as yet another example of Goethe's basic
phenomenon of color formation was not enough to satisfy Brücke's
curiosity. He insisted that the basic phenomenon was in need of a
physical explanation, since it "brought forth considerable confusion in
the heads of many." While attempting to link the visible color effects
to the microscopic properties of the solutions that caused them, Brücke
discovered an important clue. In the solution of mastic in ethanol,
which clearly showed the blue, yellow, and red colors, he could no
longer make out the individual resin particles—even at great magnification. Surely the particles had to be very small, and only reveal themselves when they accumulated in a high concentration. The same thing
happens with other turbid media: blue, yellow, and red can always be
seen whenever the particles of the solution are so small that they are

scarcely visible under the microscope. When Brücke produced a solution of lime in oxalic acid (a substance found in many plants, such as
sorrel and rhubarb), he observed a "coarse-grained crystalline precipi-
tate." In this case the colors of the basic phenomenon were barely perceptible, suggesting to Brücke that the size of the particles in the solution were decisive in producing the color effects. By experimenting
with a variety of materials, Brücke was able to demonstrate that the
color of the suspended chemical did not affect the colors seen in the
solution.

Brücke then made a giant leap from his laboratory solutions to the
earth's atmosphere. Assuming that the latter is a turbid medium, and
knowing that we see the colors of Goethe's basic phenomenon in the
sky, he deduced that the atmospheric particles "are very small and distributed throughout the atmosphere with a certain uniformity."[2]

WAVES OF LIGHT

Brücke attributed the optical properties of turbid media to the reflection and scattering of light waves. He concludes that Goethe's basic
phenomenon, as well as the colors seen in the chameleon's skin and
Earth's atmosphere, result from the separation of white sunlight into
rays of different wavelengths, and thus, different colors. Rays with
"short wavelengths" are affected by the particles causing the turbidity,
giving off a bluish light in front of a dark background. On the other
hand, when a beam of light that originally contains rays of all the colors
of the spectrum enters a turbid medium, only the red rays will pass
through. They are affected less because of their long wavelengths.

Throughout the early nineteenth century, the theory of light as a
wave phenomenon had caused a heated debate among physicists that
was just calming down at the time when Brücke carried out his studies.
Remember that Newton had surmised that light consists of particles, the
corpuscles, that propagate along straight lines. In his view, it was the "refrangibility" of a light ray that accounted for its color. However, the

nature of refrangibility remained as mysterious as that of the fits, the periodic property that he had introduced to account for the regular pattern of the colored rings. One of Newton's contemporaries, the Dutch physicist Christiaan Huygens, disagreed with these ideas and suggested instead that light propagates as an irregular sequence of shock waves. Huygens likened the light waves to sound waves and argued that they move with great speed through a medium consisting of elastic particles. As such, he deemed light not to be the actual transference of matter, as Newton had believed, but rather a tendency of motion propagating in a medium that he called the aether (a namesake of Aristotle's element of the perfect celestial spheres). There was little doubt that sound waves needed the air as a medium to travel through, as Robert Boyle had convincingly demonstrated in the 1660s. Boyle's experiments with the air pump showed that sound does not propagate in a vacuum, for the ringing of a bell that was placed in a sealed glass jar could no longer be heard when the air had been pumped out. However, the bell could still be seen—and thus he had proved that, unlike sound, light does propagate in a vacuum.

Huygens' idea of light consisting of waves had found little support until 1800, when Thomas Young took it up as an object of study. Financially independent through an inheritance left by his uncle, Young was a physician and gentleman scholar who studied topics as diverse as human color vision and Egyptian hieroglyphs. Between 1801 and 1804 he was professor of natural philosophy at the Royal Institution in London. It was here that he presented his ideas on light and color vision in public lectures.

In contemplating the nature of light, Young discarded Huygens' idea of *irregular* light waves from the outset and considered instead what happens when *periodic* light waves interact in a mechanism that he called interference. It was reasonable to assume that light stemming from different sources should interact, since obviously our environment is filled with light coming from all directions, traveling from visible objects to the eyes that see them. Light waves as Young imagined them can be likened to waves on the surface of the sea (Figure 6.3). The latter are

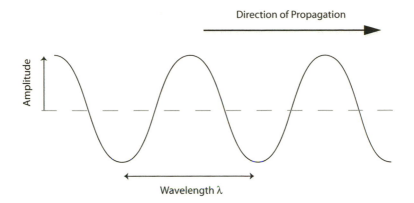

Figure 6.3 Transverse waves oscillate perpendicular to their direction of propagation.

characterized by their regular "ups" and "downs" and can be described by their height (defined as the difference between their minimum height and their maximum height) and their wavelength (defined as the distance between successive maxima or successive minima).

For Young, the analogy between waves in the sea and waves of light was a potent one. In his Bakerian lecture at the Royal Institution, delivered in November 1803, he illustrated the concept of light waves by means of optical experiments as well as simple demonstrations with a ripple tank. This was a glass vessel filled with water, in which periodic surface waves could be generated and the patterns of their propagation and deflection demonstrated by placing screens in the water.

His experiments with the ripple tank gave Young intriguing insights into the propagation of waves. Consider Figure 6.4. It shows a sketch of periodic, circular waves originating at point A and a screen with a vertical slit in it placed at some distance from this point. Up to the slit there is nothing remarkable about how the wave propagates as expanding concentric circles. However, after part of the wave front has passed the slit, it not only keeps propagating away radially from point A but seems to bend toward both sides into the zone that is out of sight as seen from A. In fact, circular wave fronts seem to originate on either edge of the

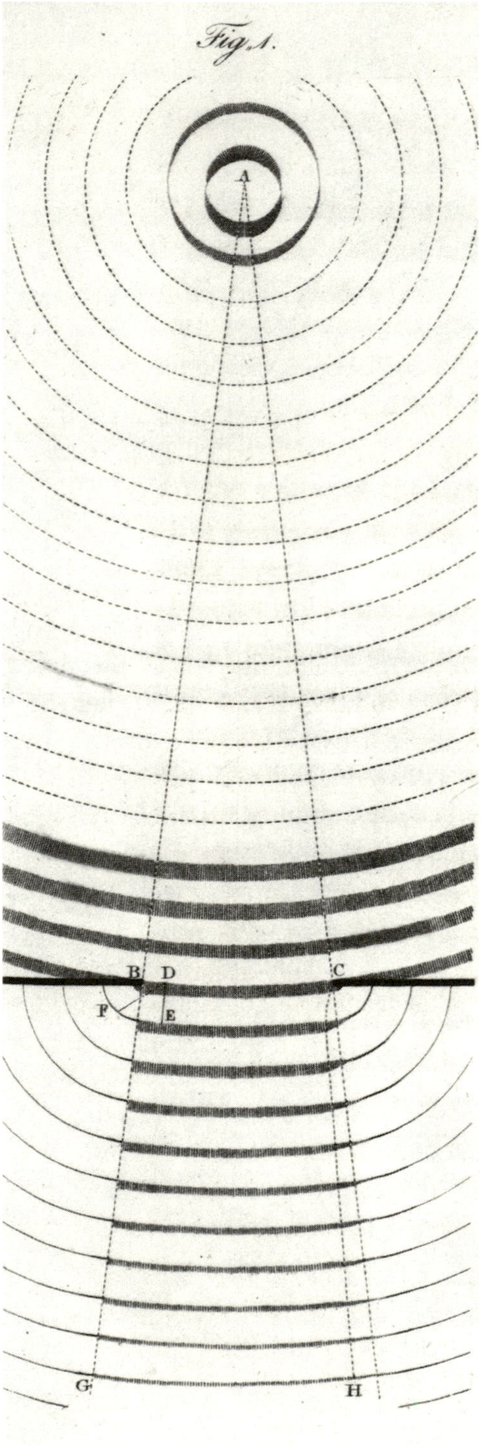

Figure 6.4 In this schematic sketch drawn from an experiment in Thomas Young's ripple tank, a circular wave originating at point A hits a screen. After part of the wave front has passed the slit, it not only keeps propagating away radially from point A but seems to bend toward both sides into a zone that is out of sight as seen from A. The waves reflected from the screen are not shown. Reproduced from Thomas Young, "On the Theory of Light and Colour," *Philosophical Transactions of the Royal Society of London* 92 (1801–1802), Figure 1 (after p. 48).

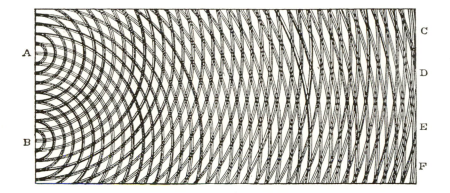

Figure 6.5 In this sketch, drawn from an experiment with the ripple tank, A and B are the origins of the circular waves, precipitating a characteristic pattern of crests and troughs. C, D, E, and F mark lines along which the crests and troughs tend to extinguish each other, while in between they tend to reinforce each other. The image is from Thomas Young, *Lectures on Natural Philosophy* (1807). Reproduced from David Park, *The Fire Within the Eye* (Princeton, N.J.: Princeton University Press, 1997), 250. Courtesy of Princeton University Press.

slit. If light is indeed a wave phenomenon, its propagation will not conform to geometrical optics, in which light rays propagate along straight lines only. The bending of waves when they pass close by opaque objects is now called diffraction. It had been familiar but mysterious to earlier scholars, who had noticed bands of light and darkness running parallel to where geometric optics would predict a sharp border between shadow and light.

Another intriguing experiment with the ripple tank was to dip two spikes into the water synchronously and periodically. Two circular wave fronts originated from these points and intersected in an intricate pattern like the one shown in Figure 6.5. Points A and B are the origins of the circular waves. At some points the maxima of these waves overlap and add up. Likewise, at those points where a minimum of one wave front meets a minimum of the other, they compound each other to form a deeper minimum. But where a maximum of one wave front meets a minimum of the other, they extinguish each other mutually (Figure 6.6). These rules of adding waves are called the principle of interference. Of

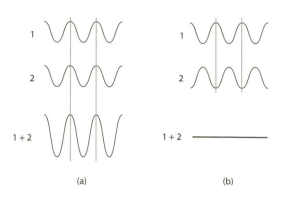

Figure 6.6 Two wave trains
with the same phase and
with equal amplitude (a)
add up to one wave train
with double the amplitude.
If the wave trains are shifted
away from each other by
half a wavelength (b), then
they mutually extinguish
each other.

course, one does not need a ripple tank to demonstrate the interference
of surface waves. Throwing pebbles into a calm pond generates the same
wave patterns, and these are familiar to almost any child.

Young suspected that light waves (which he called "undulations") behave in much the same way as the surface waves, and he proceeded to
demonstrate this conjecture by experiment. One of his setups, known
today as Young's experiment, has made it into elementary physics textbooks. In it, light from a single source shines onto a screen with two
narrow, parallel slits. When another screen is placed behind this, a distinct
pattern appears. If the incoming light is of one color, a series of alternat-
ing dark and light bands can be seen parallel to the slits. Their spacing
depends on the wavelength of light. If the incoming light is white sunlight, a complicated sequence of colors appears. Young believed that this
alleged interference of periodic light waves could explain the colors of
the mica flakes seen by Robert Hooke and the colored rings studied by
Newton. In this model, light waves of different colors would interfere at
different places. Young suspected that the wavelengths of light determine
the colors we perceive and account for the property that Newton had
termed refrangibility.

Young realized that he could calculate the wavelengths of colored light
from the spacing of the light bands that he saw on the second screen, in a
manner similar to the way Newton had determined the lengths of the fits
of refraction and reflection. It turns out that Young's light waves are four
times as long as Newton's fits. In the spectrum of visible light, violet light

TABLE 6.1
WAVELENGTHS AND COLOR NAMES OF VISIBLE LIGHT

Wavelength (nanometers)	Color Name
380–440	Violet
440–483	Ultramarine blue
483–492	Ice blue
492–542	Sea green
542–571	Leaf green
571–586	Yellow
586–610	Orange
610–705	Red

has the shortest wavelength and red light has the longest wavelength. Expressed in numbers, violet light has a wavelength of about 400 nanometers, where a nanometer (abbreviated nm) is one billionth of a meter, or one millionth of a millimeter. In other words, 2,500 violet wave peaks will fit into a space that is one millimeter long. Red light has a longer wavelength of about 650 nanometers; one millimeter will therefore accommodate about 1,500 red wave peaks. The wavelengths of the remaining colors of the spectrum, from blue to orange, fall in between (Table 6.1). Since the number of different wavelengths between violet and red is unlimited, the visible part of the spectrum must contain infinitely many different hues. The colors we designate as red, green, or blue are thus more or less arbitrary demarcations that reduce the infinite variety of colors to just a few names. There are, of course, practical advantages to this. Light waves between 440 and 492 nanometers are considered blue, and those between 492 and 571 nanometers are considered green. However, drawing the boundary between blue and green at 492 nanometers is merely a matter of definition; this number corresponds to the average person's color perceptions.

The analogies employed by Christiaan Huygens and Thomas Young in conceptualizing light raise a profound question about its nature. Sound waves, the analogy chosen by Huygens, propagate through air by inducing back-and-forth motions of the air's particles along the path of

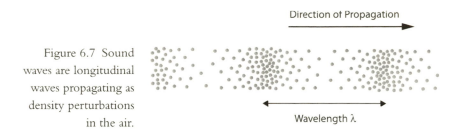

Figure 6.7 Sound waves are longitudinal waves propagating as density perturbations in the air.

propagation. Such waves are called longitudinal waves (Figure 6.7). In contrast, waves moving through a liquid, the analogy chosen by Young, oscillate perpendicular to the direction of propagation. Such waves are called transverse waves. The difference between these two types is significant, but Young's experiments do not offer any clues as to which category light waves belong to. Another optical phenomenon promised the key to unlock this mystery.

POLARIZED LIGHT

In 1669, the Danish physician and scientist Erasmus Bartholinus made a remarkable observation. Investigating the optical properties of Iceland spar, he found that this crystal separates the light that shines through it into two rays. When he laid the crystal onto a page of a book, the letters seen through the crystal were doubled. This physical curiosity came to be known as double refraction; it stood in sharp contrast to all other known optical phenomena. In particular, the laws of refraction, which had been discovered by the Dutch physicist Willebrord Snel just a few years earlier, seemed to apply to only one of the two rays. In 1690, Huygens called this ray the ordinary ray, and the other the extraordinary ray. He suspected that they were two different kinds of light that the Iceland spar could resolve somehow.

In 1808, the French physicist Étienne Louis Malus found a new clue to help solve this riddle. Malus and his wife lived in Paris across from the

Palais du Luxembourg, an eighteenth-century town palace of the French kings. One day he looked through a double-refracting tourmaline crystal at the image of the setting sun reflected in the palace's windows. Malus had expected to see two images of the sun, stemming from its ordinary and extraordinary rays, respectively. But he saw only one image. This was a puzzling observation, and Malus followed it up with an experiment that same evening. Using the same crystal, he looked from various angles at the light of a candle reflected off a pane of glass. While looking through the crystal, he discovered that there was one specific angle at which the reflected light seemed to disappear completely. The light reflected at this angle seemed to have taken on the property of the extraordinary ray known from the experiment with Iceland spar. As an adherent of Newton's corpuscle theory of light, Malus could only explain this phenomenon by assuming that a ray of light consisted of particles that are oriented or polarized in a certain way, and that differently polarized corpuscles are then separated from each other through double refraction or reflection. Malus called this phenomenon the "polarization" of light.

A contemporary of Malus', Augustin Fresnel, interpreted these observations in a completely different way. Fresnel viewed light as a wave phenomenon and saw the double refraction, as well as the polarization-through-reflection discovered by Malus, as proof that light could oscillate in discernibly different planes. The essence of both processes lay in separating the parts of a light ray that oscillate in different planes. Since only transverse waves can differ in their planes of oscillation, these observations simultaneously proved that light could not be made up of longitudinal waves but rather must consist of transverse waves.

This insight led Fresnel, at the time a young engineer employed by the French Service of Bridges and Roadways, to formulate a mechanical theory of light waves, which not only provided a consistent explanation for the polarization phenomena but also made it possible to quantitatively predict the intensity of the pattern of diffraction observed by Young, as well as of the intensity of light refracted and reflected at any given angle. In 1819 Fresnel won a prize from the French Academy of Sciences for this feat.

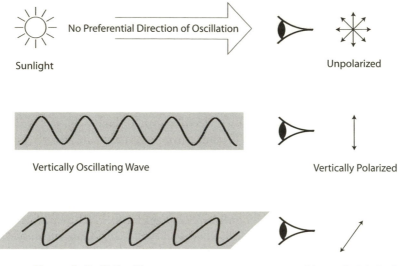

Figure 6.8 The light of the sun is unpolarized: it has no preferred plane of oscillation. In contrast, linearly polarized light oscillates in only one plane that would
seem to be inclined differently when viewed "head on" along its direction of propagation (right).

Even though it is crucial to understand basic properties of nature, polarization is an abstract concept, and its details and consequences are
somewhat difficult to comprehend. Let us consider its essentials step by
step. The wave depicted in Figure 6.3 is oscillating in the plane of the
page. It could just as well oscillate perpendicular to that, or at any other
angle. Figure 6.8 shows light oscillating in different planes. As it turns
out, sunlight has no preferred plane of oscillation. It is *unpolarized*, because in it all directions of oscillation are equally frequent. In contrast,
the wave shown in Figure 6.3 is oscillating only in the plane of the page.
In such cases, we speak of *linearly polarized light*. Completely polarized
light very seldom occurs in nature. Most often, we find unpolarized and
linearly polarized light superimposed on each other, thus combining to
form partially polarized light. Partially polarized light can be thought of
as the sum of a completely unpolarized component and a component
that is completely polarized in a specific plane (Figure 6.9).

Color Plate 1 Earth, the blue planet, seen from a distance of 37,000 kilometers (23,000 miles). Africa, Antarctica, and the Atlantic and Pacific Oceans are clearly recognizable. The photo was taken on December 7, 1972, by *Apollo 17* astronauts on their way to the moon. Courtesy of NASA.

Color Plate 2 Brightening of the blue sky near the horizon. The saturation of the blue color increases with increasing height above the horizon.

Color Plate 3 *The Heavenly Universe*, a wooden table-top painted in 1533 by Martin Schaffner, an artist from Ulm (Germany), depicts the occult correspondences among the planets, elements, humours, seasons, and colors in a fisheye perspective. The brightening of the clear sky toward the horizon is discernible. Courtesy of Staatliche Museen Kassel.

Color Plate 4 When the weather is clear, the sky is blue not only during the day (above), but also on moonlit nights (below). The moon takes the place of the sun and lights up the atmosphere. During the long exposure time required for this photograph, the apparent positions of the stars changed, such that they appear as streaks of light.

Color Plate 5 With increasing altitude, the brightness of the sky decreases until it fades into the darkness of outer space. This view was photographed from the window of a commercial aircraft at an altitude of about 10,000 meters (33,000 feet).

Color Plate 6 Colors of the rainbow.

Color Plate 7 Color of the setting sun.

Color Plate 8 The *Blue Virgin*, also called *Our Lady of the Beautiful Window* (*Notre-Dame de la Belle Verrière*), is a stained glass window in the south ambulatory of Chartres Cathedral. The window dates from the 1180s.

Color Plate 9 Brightening and bluish tint of the landscape in front of distant hills in Provence (southern France).

Color Plate 10 Leonardo da Vinci's *The Virgin of the Rocks* (c. 1506). The infant St. John (the Baptist) is seen adoring the infant Christ, accompanied by an angel. Note blueing in the distance. Courtesy of the National Gallery, London / Bridgeman Art Library.

Color Plate 11 Incense illuminated by the sun in front of a black background appears blue.

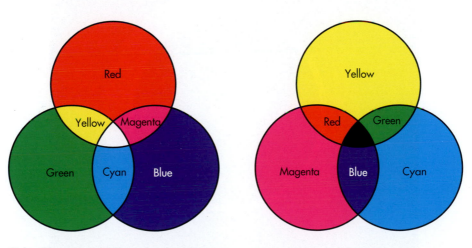

Color Plate 12 Additive color mixing.

Color Plate 13 Subtractive color mixing.

Color Plate 14 The blue of the first order is clearly visible in soap films.
If held vertically it appears at the top shortly before the film bursts.

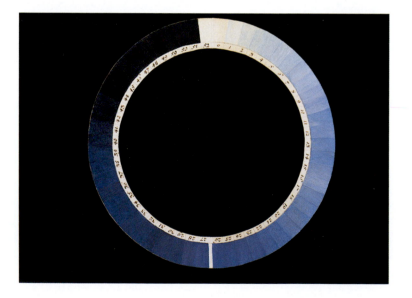

Color Plate 15 The cyanometer designed by Horace-Bénédict de
Saussure is a scale with fifty-two shades of blue for measuring the color
of the sky. Courtesy of Bibliothèque Publique et Universitaire, Geneva.

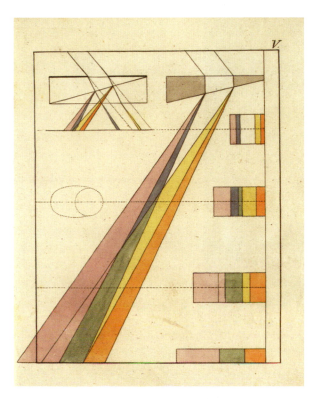

Color Plate 16 Goethe sought to replace Newton's explanation of the spectral colors with one derived from his own studies of color theory. If you look through a prism at a white area against a black background, he writes, the edge of the black area shifts over the white on one side, while on the other side the white background shifts over the black area. According to Goethe, this corresponds to the two scenarios in the basic phenomenon of color generation. He claimed that all the colors of the spectrum could be derived from it. This interpretation was roundly rejected by physicists. Reproduced from Johann Wolfgang von Goethe, *Zur Farbenlehre* (Tübingen: J. G. Cotta'sche Buchhandlung, 1810), Plate V.

Color Plate 17 Brücke's experiment on the color of turbid media is easy to replicate by dripping pistachio resin dissolved in alcohol into a glass of water. The resulting suspension, when illuminated in front of a dark background, looks bluish (right), but in front of a white background (here, a card placed behind the left-hand side of the glass) it looks slightly reddish.

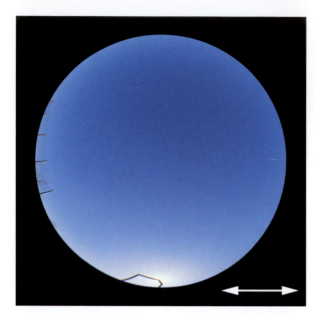

Color Plate 18 In these fisheye photographs, taken with a polarizing filter, the pattern of skylight polarization is discernible. Direct sunlight is blocked by the arc visible at the bottom. The filter transmits only light oscillating parallel to the arrows. In the upper photograph, the brightness distribution of skylight resembles that seen with the naked eye. With the filter rotated by 90 degrees (below), a dark band appears. This is because the light polarized perpendicular to the arrow is blocked by the filter.

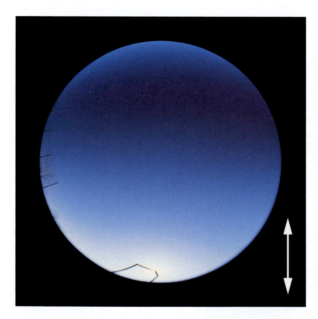

Color Plate 19 In 1870, John William Strutt used these paper disks to repeat Maxwell's experiments on color perception. This work led Strutt to ponder the origin of blue skylight.

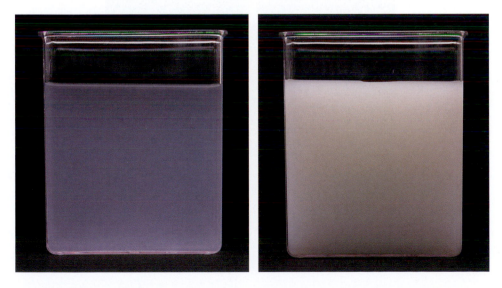

Color Plate 20 Multiple scattering can be demonstrated with a simple experiment that involves dripping milk into water. Illuminated by daylight, one drop of milk appears bluish when seen in front of a dark background (left). When more drops are added, the milk looks white (right).

Color Plate 21 At twilight, a clear sky still looks blue, even though the sun has gone down. This "blue hour" is an effect of the ozone layer in the stratosphere.

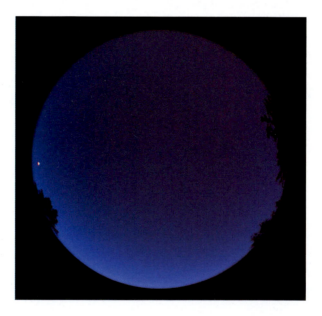

Color Plate 22 Fisheye view of the clear twilight sky. The moon is visible as a bright spot over the southern horizon (left).

Color Plate 23 Ascent of the Earth shadow in the eastern twilight sky after sunset, as seen from the summit of the Roque de los Muchachos (La Palma, Canary Islands).

Color Plate 24 Earth is not the only blue planet in the solar system. However, the blue–green color of the atmospheres of Uranus and Neptune is due to the absorption of sunlight by methane molecules. This picture of Neptune was taken by the spacecraft *Voyager 1*. Courtesy of NASA.

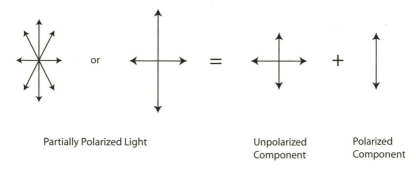

Figure 6.9 Partially polarized light can be described as the sum of an unpolarized and a completely polarized component.

This makes it easy to understand how polarizing filters work. The only light that can pass through these tools is that which is oscillating parallel to a certain plane, that is, light that is polarized in that plane (Figure 6.10). Light waves that are oscillating perpendicular to this, on the other hand, are reflected or absorbed by the filter. For light with other directions of oscillation, the filter is partly penetrable, letting only the parallel components pass through. We can therefore use the filter to determine in which direction polarized light oscillates.

Modern polarizing filters are sophisticated inventions with microscopic crystals fused into the glass. The first devices for observing polarization were simpler, based directly on the polarization-through-reflection discovered by Malus. Consider a beam of light that falls diagonally on a flat pane of glass (Figure 6.11). Part of the beam will be reflected according to the law of reflection, which holds that the angle of incidence is equal to the angle of reflection. The other part will be refracted; it changes direction and continues on its way through the glass. An everyday example of this phenomenon is a straight drinking straw that, when placed in a glass of water, looks bent. The refracted light ray is subject to Snel's law of refraction. According to this law, the angle at which the ray is refracted depends on the angle of incidence and on the ratio of the indices of refraction of glass and air. The index of refraction, to which physicists have assigned the symbol n, is a material constant of each medium through

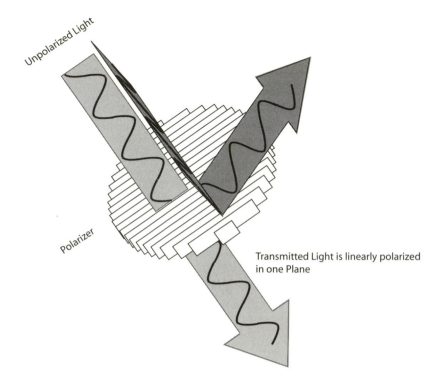

Figure 6.10 Polarizing filters let only that light through which is oscillating in a particular plane. Light waves that are oscillating perpendicular to this plane are either reflected or absorbed by the filter. Contrary to this schematic sketch, polarizing filters operate by means of polymer molecules which are oriented perpendicular to the axis of light transmission.

Figure 6.11 At a certain angle, the Brewster angle, θ_B, the reflected and the refracted light rays are perpendicular to each other. The reflected light is completely polarized in the plane that is perpendicular to this page and parallel to the reflected ray.

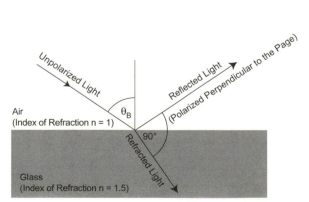

which light can propagate. By definition, a vacuum, devoid of air, has a refractive index of $n = 1$. The refractive index of air is somewhat higher, approximately $n = 1.0003$, while that of glass is about $n = 1.5$. The ratio of these indices determines how strongly a light ray is refracted when it passes from one medium into another. The index of refraction of water is about $n = 1.3$, and it follows from this that light is more weakly refracted when passing from air into water than when passing from air into glass.

At a certain angle of incidence, the reflected and refracted light rays are perpendicular to each other. This is the so-called Brewster angle, which is named after the Scottish physicist David Brewster, even though its significance had been discovered earlier by Malus. When a light ray strikes a partially reflective surface at the Brewster angle, the surface reflects light that is polarized perpendicular to the plane defined by the incident and reflected rays. The refracted light, on the other hand, is polarized parallel to this same plane. The Brewster angle depends on the refractive index of the reflecting material, because the angle of refraction is determined by this refractive index. We can therefore infer the refractive index of a body by observing how it polarizes light. Incidentally, light cannot be polarized by a metal mirror, because in that case there is no refracted light ray.

SKYLIGHT POLARIZATION

The observations that Étienne Louis Malus made in his apartment across from the Palais du Luxembourg stirred excitement among the physicists and astronomers of Paris. Soon, many of them had improvised their own devices to look for polarized light and begun to explore their surroundings with them. Some things looked the same as they did to the naked eye, and thus their light seemed to be unpolarized. Others could literally be seen in a whole new light through these new instruments. A new fad was born, and during its lifetime the light from comets' tails, red-hot metals, the moon, and atmospheric halos, as well as rainbows and some liquids, were all found to be at least partially polarized. But astronomer François Arago's 1809 discovery of the polarization of the blue light of the sky caused the biggest sensation of all. He found to his surprise that this phenomenon, considered just a few years before to be an idiosyncrasy of Iceland spar, was an everyday characteristic of daylight.

François Arago (Figure 6.12) was born in 1786 into a middle-class family from southwestern France. At first, he wanted to become an officer in the French army. But when he passed the entrance exam for the École Polytechnique in Paris, Arago decided to study physics and mathematics there. He began his studies in 1803, and just two years later he became the secretary of the Bureau des Longitudes, a scientific institution concerned with geodesy. Arago was soon sent on an expedition to Spain. This trip turned into an adventure when he was taken prisoner and detained in Spain and Algeria. Not until 1809 did he return to Paris. He was then made a professor at the École Polytechnique. During this period, he got to know Alexander von Humboldt, with whom he remained friends for the rest of his life. Having been appointed director of the Paris observatory, Arago held popular lectures on astronomy for thirty years. An active liberal democrat, he became minister of war and the navy after the revolution of 1848, and worked in this capacity to abolish slavery in the French colonies. He died in 1853.

In 1809, Arago discovered the polarization of daylight by using an apparatus that functioned on the principle of Brewster's law. To do so, he first had to overcome an obstacle: the light that is reflected at the Brewster angle is completely polarized, but only 7.5 percent of the incoming light is reflected. The remaining 92.5 percent is "lost" to refraction. Arago adopted a simple trick from Malus to increase the proportion of reflected light. When several plates of glass are laid on top of each other, the identically polarized portions of the light reflected by the individual plates add up, and the degree of polarization of the transmitted light increases also (Figure 6.13). With six parallel plates, a full 37 percent of the incoming light is polarized through reflection. All early polarizers consisted of such stacks of glass plates, and while these sufficed for the discoveries made by Arago and his colleagues, using them required some skill (Figure

Figure 6.12 François Arago in a sketch dating from the 1830s. Reproduced from Louis-Marie de la Haye de Cormenin, *Le Livre des Orateurs*, 11th ed. (Paris: Pagnerre, 1842).

6.14). Not until 1828 did William Nicol, a lecturer of natural philosophy in Edinburgh, invent a more efficient device for observing polarized light. It consists of two Iceland spar crystals that are beveled and mounted at a certain angle to each other. Nowadays it is easy to replicate Arago's observations. A (linear) polarizing filter is all you need, and they can be found in any photography shop.

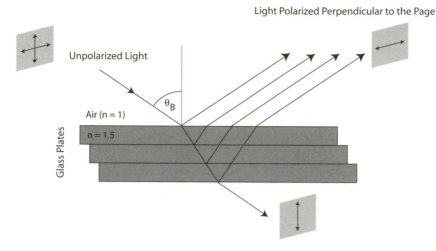

Figure 6.13 When several plates of glass are stacked on top of each other, the proportion of light that is reflected increases. At the same time, the intensity of the light refracted in the glass decreases, but its degree of polarization increases.

Figure 6.14 Malus' polariscope uses two stacks of glass plates to separate the two perpendicular planes of oscillation in an incident light beam from each other. The polariscope is constructed symmetrically; this makes it possible to see the polarization in the original direction of a light ray. The transmitted light is polarized in the plane of the page.

As mentioned above, you can infer that the light coming from a certain direction is polarized if, on looking through a polarizing filter, its intensity varies appreciably as you rotate the filter clockwise or counterclockwise. If you look at a clear daytime sky in this way, you can easily tell that its light is polarized. You may even discover that skylight is

polarized in a distinct pattern, something that Arago had already noticed in his time. The polarization of daylight is most pronounced at a
90-degree angle from the sun, in just that zone where the blue color is
at its darkest and most intense. In areas closer to or farther from the
sun, the degree of polarization decreases. In these directions the blue
also grows less intense. More recent observations have revealed the pattern of this polarization. The degree of polarization is 75 percent at a
right angle to the sun and diminishes at points closer to or farther from
the sun. On average, 40 percent of all daylight is polarized. With our
simple polarizing filter, we can tell that even the maximal polarization
found in the sky is not complete, that is, it is less than 100 percent: if it
were complete, then there would be a certain angle at which the filter
would let no light through at all, and we would see complete darkness.
But that is not the case.

In addition to the degree of polarization, its orientation also varies in
a characteristic manner (Color Plate 18). We can imagine that every
point in the sky lies on the perimeter of a circle whose center marks the
direction to the sun (Figure 6.15, left). The polarization planes at such
points nestle themselves up against these imaginary circles like tangents.
At a given point the polarization is thus perpendicular to the plane defined by this point, the sun, and the eye of the beholder. Skylight is *tangentially* polarized. In its daily arc of apparent motion across the sky, the
sun pulls this pattern of polarization along with it. When the sun is in
the south at noon, the area of greatest polarization shifts northward. In
the twilight just after sundown or before sunrise, the region of greatest
polarization at an angular distance of 90 degrees from the sun passes
through the zenith (Figure 6.15, right). This is the grand scheme of
clear-sky polarization, but its details are a little more complex. Once
again it was Arago who noticed this, discovering that about 20 degrees
above the antisolar direction, at sunset over the eastern horizon, the direction of polarization changes around the so-called Arago point. The
polarization vanishes in this direction. The polarization is vertical above
this point, but horizontal below it. By 1840, Jacques Babinet had discovered another neutral point, one that is located about 20 degrees above
the sun but is difficult to discern owing to the overpowering light of the

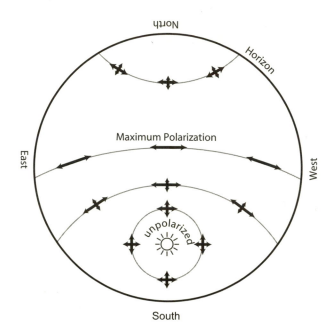

Figure 6.15 Projections of the visible sky onto a circular disk with the zenith in the center and the horizon along the circumference. This diagram is best understood by holding it up to the sky, aligning the cardinal directions in the diagram with those in the environment. The picture shows how the sky's pattern of polarization follows along with the sun over the course of a day. The greatest polarization is always found at a 90-degree angle from the sun. The polarization decreases closer to the sun, as well as near the antisolar direction. On the left is the polarization pattern at noon, when the sun is in the south. On the right is the polarization at dusk; now the sun is at the western horizon. The Arago point (A) is above the eastern horizon.

sun. David Brewster deduced that there must be another neutral point the same distance *below* the sun, which he actually succeeded in observing soon afterward. These neutral points are called the Babinet point and the Brewster point in honor of their discoverers.[3]

The history of its discovery suggests that polarized light is a phenomenon that is imperceptible to the naked eye and requires instruments in

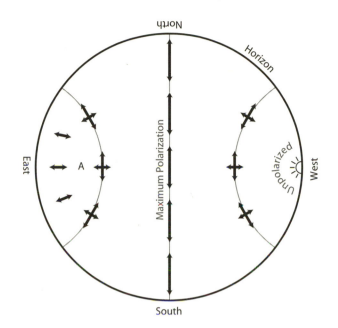

Figure 6.15 (*continued*)

order to be observed. That is not generally true, or at least not for every observer. If you look at those parts of a clear blue sky that are most strongly polarized (in general, at a right angle to the sun), you may notice a faint pattern of two spots that look darker than the sky around them, with a yellowish bar in between. This remarkable phenomenon was discovered by the Austrian mineralogist Wilhelm Haidinger in 1846. Known as Haidinger's brush, its pattern is sometimes likened to a butterfly. A small butterfly, one might add, for it measures only 5 degrees across—that is, about ten times the size of the full moon or the angle subtended at the eye by two fingers extended at arm's length. It is reported that James Clerk Maxwell, Hermann von Helmholtz, and even the Russian writer Leo Tolstoy of *War and Peace* fame have seen this phenomenon. Apparently, it took Helmholtz twelve years of practice to see it. Sunset or sunrise seem to be the best times for beginners; at these times the brush should be most clearly visible toward the zenith.

THE PARADOX OF POLARIZATION

Arago noticed early on that the extent of the polarization is subject to large fluctuations. A cirrus haze weakens the polarization considerably, and in overcast skies it is altogether indiscernible. A few years after the discovery of skylight polarization, a growing consensus held that its cause is the same as that of the blue color. All observations of deep blue skies revealed them to be strongly polarized, while milky, clouded skies proved only slightly polarized. In the 1840s, the English astronomer John Herschel wrote:

> The more the subject [the polarization of daylight] is considered, the more it will be found beset with difficulties; and its explanation, when arrived at, will probably be found to carry with it that of the blue colour of the sky itself and of the great quantity of light it actually does send down to us. We may observe, too, that it is only where the purity of the sky is most absolute that the polarization is developed in its highest degree, and that where there is the slightest perceptible tendency to cirrus it is materially impaired.[4]

It was rather difficult to tie in an explanation of the sky's blue color with that of its polarization. Consider Newton's explanation, which remained popular at the beginning of the nineteenth century, even though a growing shadow of doubt was cast on it. He had assumed that small water droplets or ice crystals, which were floating in the air, separated the blue and violet light rays out of the white composite light of the sun, distributing them throughout the sky. At the beginning of the nineteenth century, this explanation was the only theory that predicted the location of maximal polarization; all the other explanations either spoke in vague terms about the separation of different-colored light rays in the atmosphere or assumed that the air particles were blue in color. This, however, does not account for the polarization of skylight.

The centerpiece of the Newtonian explanation was thus the reflection of light rays, and so the Brewster angle should predict in which direction

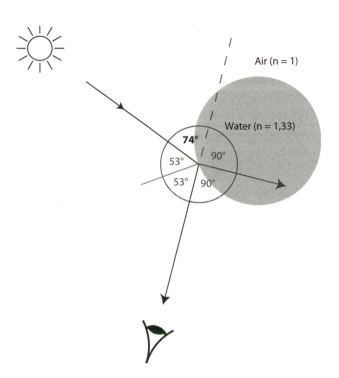

Figure 6.16 Newton's explanation of the blue sky makes it possible to calculate the expected maximal polarization according to Brewster's law. In this case we will assume that small water droplets floating in the air reflect the sunlight. This gives us a maximal polarization at an angle of 74 degrees from the sun—a value that is impossible to reconcile with the observed maximum at a 90-degree angle from the sun.

the maximal polarization would be found. The refractive indices of water and ice (about $n = 1.33$), give us a Brewster angle of about 53 degrees. As we can infer from Figure 6.16, this value predicts that maximal polarization of skylight should be seen at an angular distance of 74 degrees from the sun. This angle seemed to agree with the earliest observations, according to which the polarization appeared greatest at *approximately* a right angle to the sun. The location of the maximum could not be determined more precisely than this at first, because the polarization is continuously distributed across the entire sky, so that clear differences can only be detected over large angular distances.

Although the earliest observations of skylight polarization appeared to confirm Newton's theory, which correctly predicted the pattern of tangential polarization, it did not take long before further observations began to discredit it. In the 1820s, John Herschel and his colleagues pinpointed the polarization maximum more precisely and found that it is located at an angle of exactly 90 degrees from the sun. The 74 degrees predicted by the Newtonian theory was clearly out of the question.

Given that maximum polarization is observed at a 90-degree angle from the sun, we can invert the above calculation and, assuming that the polarization of skylight is caused by the reflection of sunlight on droplets in the atmosphere, infer the refractive index of the matter they are made up of. This rather novel approach to investigating the nature of matter floating in the sky was exactly what Herschel tried to do—and he came up against a paradox. A polarization maximum at 90 degrees corresponds to a polarization angle of 45 degrees. When Brewster's formula is applied, it turns out that the refractive index of the reflective bodies must be about $n = 1$. Meanwhile, air has a refractive index of approximately $n = 1.0003$, in contrast to the much higher refractive indices of liquids and solid bodies. Thus, if the blue light of the clear sky is the result of reflections, then they must take place at a juncture from air to air, as Herschel wrote. However, between two media of equal refractive index there can be no reflection, because under those circumstances a light ray will propagate along a straight line without changing directions. The propagation of light through media with differing refractive indices is, after all, the precondition for its refraction and reflection. The polarization of skylight thus shows that reflections of sunlight off atmospheric particles cannot explain the sky's color. This was a conclusion that John Herschel did not want to accept, even though his observations and calculations all pointed in that direction. Contemporary physics was at its wits' end, and it is no wonder that the color and polarization of the sky remained for John Herschel "the two great standing enigmas of meteorology" of his time.[5]

SKY IN A TEST TUBE

Ernst Wilhelm von Brücke had claimed that, in his studies of turbid media, he had brought the colors of the sky into his laboratory. But it would take measurements of the polarization to show whether Brücke had in fact recreated blue skylight in a test tube. After all, Arago's and Herschel's observations had revealed a close relationship between the sky's color and its polarization. Did the same apply to turbid media in the laboratory? The Irish physicist John Tyndall set out to investigate this issue.

Tyndall (Figure 6.17), born in 1820, worked as a surveyor in England before he began his academic career by enrolling at the University of Marburg in Germany. He earned a doctoral degree in mathematics and then turned his attention to experimental physics. In 1853, two years after returning to England, Tyndall became the professor of natural philosophy at the Royal Institution in London, and held this position until 1887. He died in 1893.

Tyndall had first studied the magnetism of crystals and how alpine glaciers flow. The fact that he took an interest in such disparate subjects can be explained by his great passion for mountain climbing. In fact, during his summer vacations, Tyndall became a pioneer of alpine mountain climbing. He fell just a little bit short of becoming the first to climb the Matterhorn near Zermatt in Switzerland, and he succeeded conquering Mont Blanc in a solo climb. Around 1860, Tyndall began to study the effects of light and heat rays on gases and vapors. While at it, he discovered the greenhouse effect of water vapor and carbon dioxide gas and found out that light and heat can catalyze chemical reactions in gases. Tyndall wrote that it is possible to destroy airborne microorganisms with heat radiation, and suggested using this type of radiation to kill germs floating in the air.

In the fall of 1868, Tyndall began to investigate the chemical effects of light rays on gases in his laboratory on Albemarle Street in London. The apparatus he used for this project is depicted in Figure 6.18. Its center-

Figure 6.17 John Tyndall in a photograph taken in 1857. Courtesy of The Royal Institution, London / Bridgeman Art Library.

piece is the experimentation pipe, a glass tube about one meter in length and eight centimeters in diameter, which he could fill with different gases and vapors for observation. Before each experiment he cleaned the pipe thoroughly, pumped the air out of it, and sealed it off with a valve. Then he filled a test tube with the liquids to be investigated and con-

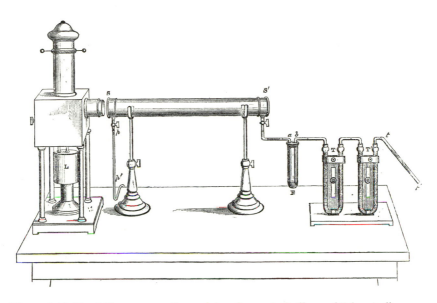

Figure 6.18 Tyndall's apparatus for studying the optical effects of "chemically active clouds." In the middle is the experimentation pipe made of glass (SS'), into which he could draw the vapors of the test substance (E). The experimentation pipe was illuminated by an electric lamp (L). Reproduced from John Tyndall, "On the Action of Rays of High Refrangibility upon Gaseous Matter," *Philosophical Transactions of the Royal Society of London* 160 (1870): 337.

nected it to the experimentation pipe with a hose. Using another hose, the test tube was then connected to a device designed to purify the air as it flowed through. The experiments began with the opening of the valve on the experimentation pipe, which was largely devoid of air. Owing to the low pressure, the pipe sucked in air, which first had to pass through the liquid in the test tube, thus filling the glass pipe with a mixture of air and the vapor of the liquid. In his completely darkened laboratory, Tyndall used an electric lamp to illuminate the vapors in the pipe from the side in order to study their appearance, as well as the color and polarization of the light they emitted.

On the morning of October 10, 1868, Tyndall drew a mixture of air and hydrochloric acid vapor into his glass pipe. At first this gaseous mixture was invisible. But over time, a cloud condensed in the pipe.

When illuminated with white light, it appeared at first violet, then gradually took on a bluish and finally a white color. He wrote in his notebook, *"connect this blue with the colour of the sky."*[6] Tyndall was familiar with reports of skylight polarization, so he used a Nicol prism to test the polarization of the light given off by the hydrochloric acid vapor. Just as the polarization of the sky is most pronounced at a 90-degree angle from the sun, the light given off by the hydrochloric acid vapor was completely polarized at a right angle to the electric lamp, but was only partially polarized at every other angle. Tyndall could discern the polarization only as long as the light continued to produce a blue color in the hydrochloric acid vapor. But by the time the vapor began to appear white, the polarization had disappeared, together with the blue color.

Tyndall knew of Brücke's experiments with mastic dissolved in alcohol and water, but notes that the color given off by the vapors he investigated resembled the sky's hue much more closely:

> The blue, moreover, is purer and more skylike than that obtained from Brücke's turbid medium. There could scarcely be a more impressive illustration of Newton's mode of regarding the generation of the color of the firmament than that here exhibited; for never, even in the skies of the Alps, have I seen a richer or purer blue than that attainable by a suitable disposition of the light falling upon the precipitated vapor. May not the aequeous vapor of our atmosphere act in a similar manner?[7]

Note that Tyndall considers his experiments to be an illustration of "Newton's mode" of producing colors. And his suspicion that "aequeous vapor" is the agent of color production in the atmosphere comes close to Newton's argument in *Opticks*. Tyndall goes one step further and insists that he has created a piece of artificial sky in his laboratory:

> The incipient actinic clouds are to all intents and purposes pieces of artificial sky, and they furnish an experimental demonstration of the constitution of the real one.[8]

Without question, Tyndall's results were impressive. But the air surely does not consist of hydrochloric acid vapor any more than it contains Brücke's mastic grains, and Tyndall was aware of this. The evidence merely pointed to the fact that the particles of hydrochloric acid vapor have similar optical properties to the particles of the air. Their small sizes seemed to be the most likely basis for this commonality, and Tyndall's future experiments would bear this suspicion out: the longer the vapor remained in the glass pipe, the more it appeared to condense and form larger particles. In each case the blue disappeared.

The colors and the polarization that Tyndall observed are not confined to hydrochloric acid vapor. He patiently repeated the experiment with many different substances. In each case, the blue color of the vapor particles was associated with maximal polarization at a 90-degree angle to the lamp, while the fading of the blue color was always accompanied by a decrease in polarization. Table 6.2 shows Tyndall's results for five substances that have significantly different refractive indices in their liquid states. According to Brewster's law, the angles of greatest polarization should also differ significantly between these substances, provided the polarization was due to the light being reflected off the vapor particles. And yet in no case was the observed polarization maximum found in the direction calculated using Brewster's law. Rather, the angle of greatest polarization appeared to be independent of the index of refraction. Moreover, he realized that the blue color seen in the illuminated vapors does not depend on the color of the liquid. Therefore, the only thing the particles have in common is their small size. It is the sole factor that makes the different vapors in the glass pipe behave optically like the microscopic particles in the cloud-free atmosphere.

To summarize, small airborne particles can polarize light at a right angle to a light source and cause it to appear blue from the side. This well-supported finding of Tyndall's was quickly met with a positive response from his contemporaries. The color and polarization phenomena that he observed are still known as the Tyndall effect to this day. But its theoretical interpretation posed a problem, because his findings could not be explained by geometric optics. Tyndall had tried to explain the polarization

TABLE 6.2

SOME OF TYNDALL'S OBSERVATIONS OF "CHEMICALLY ACTIVE CLOUDS" MADE UP OF DIFFERENT SUBSTANCES IN AUTUMN 1868

Substance	Color in Liquid State	Index of Refraction of Liquid	Color when Laterally Illuminated by White Light	Observed Angle of Maximum Polarization	Maximum Polarization According to Brewster's Formula (degrees)
Nitrite of butyl (1-nitrobutan)	Transparent, yellowish	1.42	"A very white and brilliant cloud"		70.3
Nitrite of amyl (1-nitropentan)	Transparent, yellowish	1.43	"Tinged throughout with iridescent colours"	Almost nil	70.0
Toluol	Transparent, colorless	1.52	"Blue as pure as that of an ordinary cloudless sky in England"	Complete at 90 degrees	66.9
Benzol	Transparent, colorless	1.53	"An exquisite sky-blue colour"	Complete at 90 degrees	66.6
Carbon disulfide	Transparent, colorless	1.67	Blue	Complete at exactly 90 degrees	61.7

SOURCE: John Tyndall, "On the Action of Rays of high Refrangibility upon Gaseous Matter," *Philosophical Transactions of the Royal Society of London* 160 (1870): 333–65.

on the basis of Brewster's law within the framework of geometric optics, and the color as a consequence of the wave nature of light. This was a contradiction that he himself could not resolve. While Brewster's law had until then proved applicable only to very large bodies, this explanation of the blue color presupposed them to be very small. Surely the polarization would also have to be understood as the effect of small particles on light *waves*?

The notion that geometric optics may not be valid with respect to small particles had already occurred to Tyndall's friend Rudolf Clausius. The latter had tried, in 1853, to salvage Newton's explanation of the sky's blue color by proposing that it is caused by the reflection of sunlight off thin-skinned water bubbles. However, he was cautious and emphasized that his theory could only have validity insofar as the known laws of geometric optics applied to such small particles:

> But if we assume that the particles which are active in the atmosphere are so small that those laws no longer apply to them, then these conclusions are also invalid. In that case, however, [Newton's] theory concerning the colors of thin wafers would no longer be applicable either, and a new development would be needed, whereby it will be particularly important to take into account the extent to which this assumption can be reconciled with the polarization of the light coming from the sky.[9]

THE SKY IN THE EYE

Shortly after World War II, the ethologist Karl von Frisch discovered that the polarization of the sky is not only of interest to physicists. Bees use it every day, navigating by the polarization pattern of skylight. This is made possible by cells in the eyes of honeybees that are sensitive to polarized light in the ultraviolet range of the spectrum.

A bee's compound eye consists of about 5,000 units called ommatidia. Each individual ommatidium has its own lens, and below this are light-sensitive visual cells that contain rhodopsin molecules, a specific

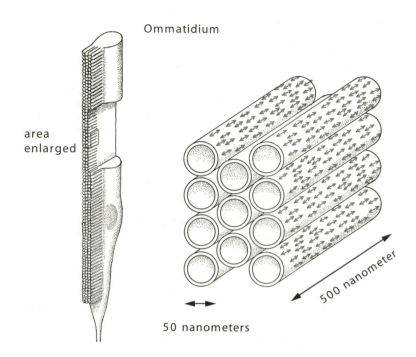

Figure 6.19 In the ommatidia, the elements of bees' eyes, polarization-sensitive rhodopsin molecules are arranged parallel to each other. This allows the bees to perceive the pattern of skylight polarization in one glimpse and to use this information for navigating. Reproduced from Rüdiger Wehner, "Polarized-Light Navigation by Insects," *Scientific American* 235, no. 1 (July 1976): 110. Courtesy of Scientific American.

kind of light-absorbing pigment. Unlike the eyes of vertebrates, the rhodopsin molecules in a bee's eye are arranged parallel to the long axis of the visual cells (Figure 6.19). This parallel alignment enables the bees to sense polarized light. Only those cells that are sensitive to ultraviolet light are able to perceive the polarization. Thus, they can clearly see the polarization of skylight, which is especially pronounced in the ultraviolet range of the spectrum. On the other hand, those cells in the bees' visual apparatus that play an important role in perceiving the colors of flowers are polarization-blind. Since the ommatidia only perceive the light that is polarized in the direction of their visual cells, but are at the same time pointing in many different directions, bees can take in the polarization pattern of the entire sky at a glance.

So, if bees see a portion of the blue sky, all they have to do is rotate around their vertical axis until the polarization of the sky matches up with the pattern imprinted in their eyes. This is very useful in comparison to navigating by the sun, since all it takes is one cloud to blot out the sun. If bees navigated by the sun, then even a thin cloud cover could make them lose their way. In contrast, the characteristic pattern of polarization is spread out across the entire sky and is affected only where it is obscured by clouds. Even in a partly overcast sky, the polarization pattern remains unchanged in those areas that are free of clouds.

Polarization only becomes a useful means of navigation for bees when combined with their internal clock, since the pattern of polarization is determined by the position of the sun. When the bees return to the hive, they use a "dance language" to tell their comrades which way to fly, in relation to the polarization, to find the richest food sources. Karl von Frisch was awarded the 1973 Nobel prize in physiology or medicine for this fascinating discovery dating from the 1930s.

In the 1920s, Felix Santschi of Switzerland had already discovered that some ant species use the sky's polarization to navigate and bring their quarry home safely. Santschi worked as a doctor in the Tunisian city of Kairouan but devoted his free time to the study of desert ants. Since the beginning of the 1970s, his countryman Rüdiger Wehner of the Zoological Institute at the University of Zurich has been studying desert ants of the genus *Cataglyphis* in the Tunisian Sahara. Wehner and his staff have discovered that these animals also carry around a simplified map of the sky's polarization pattern in their brains. Although this map corresponds to the polarization pattern at sunrise and sunset, even at noon it is accurate enough to enable *Cataglyphis* to make a "beeline" for its mound after having zigged and zagged hundreds of meters away from it. But when the sky's polarization pattern is obscured by even a light cloud cover, the ants will venture forth from their mound only for very short hunting forays.

In the late 1960s, the Danish archaeologist Thorkild Ramskou put forward the conjecture that the Vikings may also have used the sky's polarization as a navigation aid. He had gotten the idea from the mention of *sunstones* in Norse sagas. It was thought that this might refer to feldspar crystals, which can function as polarizers. On the face of it, this seems plausible and practical, since the Vikings' sea voyages led them through parts of the north Atlantic that are often overcast, which makes it impossible to navigate by the sun.

However, a look at the archaeological evidence and the physical reality suffices to disabuse us of this notion. To this day, no excavation has ever turned up a crystal that could serve as a polarizer. Moreover, navigating with such a crystal is quite difficult because the pattern of polarization changes with the position of the sun, which in turn depends in a complicated way on the geographical latitude. Catching a brief glimpse of the polarization in a break in the cloud cover gives you very little indication as to the geographical latitude. The eyes of bees and ants, which can quickly take in the entire pattern of polarization, are vastly superior to the hypothetical sunstones of the Vikings. The sunstone as a navigation tool is probably only a legend. Presumably the Vikings used wind and wave directions, as well as the sun, for orientation.

CHAPTER 7

Lord Rayleigh's Scattering

How do we see the blue sky? With our eyes, of course. But how do our eyes perceive color? This question had long tantalized curious scientists, but in the nineteenth century many of them were convinced that they were making progress toward an answer. The physiological colors studied by Goethe suggested that the perceptual apparatus, including the brain, was involved in color vision. But even if this was true, there needed to be some receptors in the eye, most likely in the retina, that were sensitive to different hues, that is, light of different refrangibilities or wavelengths. In 1802, Thomas Young felt that the goal of identifying the properties of these color receptors was within reach. He considered the color spectrum to be continuous, which means that there is an infinite number of different-colored rays. On the other hand, he was convinced that there could only be a finite number of types of color receptors in the retina. Young believed that nature is parsimonious and that therefore the eyes would be equipped with as few types of receptors as possible, even if each type existed in large numbers in the retina. Painters were known to produce all the colors of the spectrum by mixing red, yellow, and blue pigments in suitable proportions. This is subtractive color mixing. Likewise, it had been known since Newton's days that blue, green, and red light rays, when mixed in the correct proportions, can produce light of any arbitrary spectral color. This is called additive color mixing. Young concluded that there must be three types of receptors, and that these were sensitive to red, green, and blue light, respectively. Half a century later, Hermann von Helmholtz reformulated this hypothesis as the so-called Young-Helmholtz theory of color vision. Among the many fascinating features of this theory is a simple explanation for partial color blindness, which it blames on the failure of one or more types of color receptors.

In 1849, the eighteen-year-old Scotsman James Clerk Maxwell (Figure 7.1) wondered if it was possible to gain insights into the functioning of our visual perception with simple experiments that did not make any reference to anatomy. He began to develop more refined methods for studying this question. Maxwell's apparatus, the color top, consisted of a set of round paper disks in the colors red, yellow, green, and blue, as well as black and white. These he attached to a wooden top that could be spun rapidly with a crank handle. Maxwell constructed the paper disks so that they could be fitted together in arbitrary proportions. When the top was spun, the color sections seemed to melt into a new color that was the additive mixture of its components. Maxwell proceeded to arrange his colored papers into a larger outer disk and a smaller inner disk. When these were rotated at high speeds, two different mixtures could be compared. By shifting the proportions of the blue, green, and red sections, Maxwell was able to mix any arbitrary spectral color—providing clear support for Young's hypothesis. By carefully noting the proportions of a set of colors that yielded a new hue, he was able to work out a system of "color equations" for additive color mixing. Unlike Newton, who had portrayed the set of primary colors as a circle, Maxwell placed them in a triangle with red, green, and blue at the corners. His work set the stage for scientific colorimetry.

Maxwell's work on color vision culminated in a lecture, "On the Theory of the Three Primary Colours," that he delivered at London's Royal Institution on May 17, 1861.[1] Building on Young's conceptual theory of the three types of color receptors, Maxwell showed photographs of a colorful ribbon that he had taken through glass vessels containing red, green, and blue solutions, respectively. Each of the photographs, when viewed through the appropriate solution, was an image that would be visible to one of Young's color receptors. Maxwell then proceeded to superimpose the three photographs, and amazingly, the resulting composite image closely resembled the ribbon's original colors. The audience was baffled. This young man had invented color photography.

The riddle of the color receptors was not resolved until 1964, when a group of American physiologists was able to prove with microscopic

Figure 7.1 Photograph of James Clerk Maxwell from 1851. The twenty-year-old is holding a color top, which he used to conduct experiments on color perception. Courtesy of University Library, Cambridge.

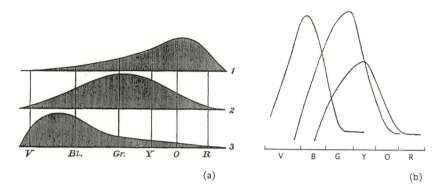

Figure 7.2 Wavelength dependence of the spectral sensitivity of the three types of cones in the human eye, as conjectured by Hermann von Helmholtz in the mid-nineteenth century (a) and as measured by P. K. Brown and G. Wald in 1964 (b). Reproduced from David Park, *The Fire Within the Eye* (Princeton, N.J.: Princeton University Press, 1997), 304. Courtesy of Princeton University Press.

studies that there are indeed three kinds of cone-shaped receptors in our retinas that are sensitive to blue, green, and red light—a belated triumph for Young's hypothesis (Figure 7.2). The color receptors in the retina are now called cones. Our eyes also contain another kind of light-sensitive cells called rods, which, though more sensitive to light, are insensitive to color.

Nine years after Maxwell's lectures at the Royal Institution, in early July 1870, the young physicist John William Strutt set out to replicate the Scottman's early experiments with color. Strutt began by constructing a color top and repeating the observations in order to determine his own set of coefficients for the color equations (Color Plate 19). His measurements corroborated the Scotsman's findings. However, when Strutt (Figure 7.3) repeated the experiment on July 23, he got a surprise. Although the color equations gave a consistent set of coefficients, he realized that to mix some colors he had to use a larger proportion of the red disk. Upon going over his notes, he immediately understood what had happened. All previous trials had taken place under a cloud-covered sky, but on July 23, there was an "unusually blue sky which sometimes accompanies a high wind."[2] Thus the color equations brought to light

Figure 7.3 John William Strutt, photographic self-portrait taken in 1870. Reproduced from Robert John Strutt, *Life of John William Strutt, Third Baron Rayleigh*
(London: Edward Arnold & Co., 1924).

what had been hidden to Strutt's immediate perception: the blue light of the sky influences how we see colors in the environment. This finding provided additional evidence of color constancy and underscored the deficit of red rays in skylight. Though perhaps not a remarkable finding in itself, it set Strutt on the track to take a fresh look at one of nature's most enduring riddles.

In 1842, Strutt was born into the landed British gentry at his family's estate in the Essex countryside, fifty miles east of London. In 1821 his grandmother, Charlotte Mary Strutt, was made the first Baroness Rayleigh. Her husband, Joseph Holden Strutt, had earned military honors fighting Napoleon, and thus King George III had offered to raise him to the peerage. However, Joseph declined this honor, preferring to pursue his political career as a Member of Parliament. With Charlotte Mary becoming the first baroness, the couple was entitled to secure Terling Place, an eighteenth-century country house near Chelmsford, from the crown. The mansion came with a gift of land. Rented out to tenants, it mostly consisted of cattle pasture. By the late nineteenth century, the Rayleighs were among the largest milk producers in England. As the eldest son of the second lord, John William was destined to become the third Lord Rayleigh. But in July 1870 this was yet to happen, and for now his privileged family background gave him the freedom to study whichever subjects he wished. Strutt chose optics and acoustics, and the investigation of color vision became one of his first projects. Soon a wing of the mansion at Terling was transformed into a physics laboratory.

At Cambridge University, John William had earned the highest score on the 1865 mathematics final examination and was awarded the title of that year's Senior Wrangler. Cambridge University was the top training ground for Britain's elite in mathematics and the natural sciences. It was a competitive environment, with students being drilled by their tutors. Students were judged mostly by their performance on the final exams, and their career prospects hinged on this evaluation. The list of the top graduates was published annually in local newspapers as well as the university bulletin, and this was the basis on which fellowships and even professorships were awarded. As Senior Wrangler, Strutt was on the fast track to a prestigious career. Finishing with distinction in 1866, the twenty-five-year-old became a fellow of Trinity College.

THE PROPER MEANING OF "REFLECTION"

Strutt's surprising observation of July 23, 1870, aroused his curiosity about the uneven spectral composition of skylight—as yet an unsolved problem, as he learned from an article recently published by John Tyndall in the *Philosophical Magazine*, the leading British journal of physics at the time. Tyndall had made it clear that its explanation would also need to account for the polarization of skylight. Strutt writes:

> It is now, I believe, generally admitted that the light which we re-
> ceive from the clear sky is due in one way or another to small sus-
> pended particles which divert the light from its regular course. On
> this point the experiments of Tyndall with precipitated clouds
> seem quite decisive. Whenever the particles of the foreign matter
> are sufficiently fine, the light emitted laterally is blue in color and,
> in a direction perpendicular to that of the incident beam, is *com-
> pletely polarized.*[3]

Strutt held Tyndall's skills as an experimental scientist in high esteem and he concurred with the conclusion that the light emitted laterally from the precipitated clouds in the experimental tube consisted mostly of short wavelengths and was almost perfectly polarized. Yet Strutt also understood that Tyndall, in applying Brewster's law of polarization to interpret his findings, had applied a law of geometric optics that had been verified only for refractive surfaces much larger than the suspended particles. Only particles much larger than the wavelength of the illuminating light are able to *reflect* this light. Strutt realized that on microscopic scales, the word *reflection* loses its quotidian meaning. Consequently, a valid explanation of the polarization of skylight should be based strictly on the wave properties of light and renounce the use of Brewster's law.

Realizing that Tyndall had misconceived the meaning of reflection with regard to light waves hitting small particles, Strutt set out to explore the interaction of light and small particles through a strict application of the wave theory. He was well-positioned for such an endeavour. At Cambridge, some of his teachers and colleagues had studied wave phenomena. Back in the 1850s, Strutt's optics teacher George Gabriel Stokes had investigated the propagation and polarization of mechanical transverse waves in the aether. This still hypothetical medium was thought to make the propagation of light possible. Maxwell, who as an examiner of the university spent much time in Cambridge, had proposed a new theory of electromagnetism in 1864 and advanced the conjecture that light consists of transverse electromagnetic waves. Strutt was in touch with both of these scholars and possessed the necessary mathematical and intuitive skills to tackle this difficult problem.

In the autumn of 1870, after pondering the problem for four months, Strutt succeeded in finding a theory that could account for the essential properties of blue skylight, as well as the light observed in Tyndall's experiments. It describes the *scattering* of light by small particles—that is, how the latter are set in oscillation by an incoming light wave and re-emit the received energy as waves in all directions. This mechanism presupposes light to consist of transverse waves, a fact well established by previous observations of its polarization. While Ernst Wilhelm von Brücke and John Tyndall had written of light that was "scattered," Strutt was the first to say precisely what he meant by this. He published his results in a paper, "On the Light from the Sky, Its Colour and Polarization," in the *Philosophical Magazine's* February 1871 issue. This article marks the birth of our modern understanding of the sky's blue color and its polarization.

In his paper, Strutt is eager to give a clear exposition of the physics underlying his mathematical derivations, and it is this style that makes the article worthwhile reading even today. Rather than delving straight into complicated mathematics, Strutt begins with intuitive physical reasoning in a thought experiment: Given a small particle suspended in an elastic medium, how does it react when hit by an incoming mechanical transverse wave? Strutt was convinced that this model describes the

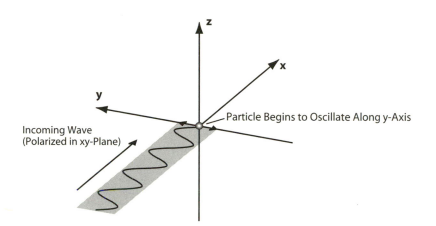

Figure 7.4 A spherical particle is induced into oscillating by an incoming, transverse wave polarized in the xy-plane.

basic conditions in the atmosphere. After all, observations of polarization had shown that light was a transverse wave, and moreover, Tyndall's experiments had clearly demonstrated that small particles suspended in the air were capable of deflecting light in a manner that closely resembled the light from the sky. The elastic medium, on the other hand, was meant to correspond to the aether as the medium of light propagation. Hypothetical as it was, Strutt sided with most contemporary physicists in assuming its existence.

A THEORY OF LIGHT SCATTERING

Let us begin by considering Figure 7.4. It shows a three-dimensional Cartesian coordinate system, that is, a coordinate system with its x-, y-, and z-axes perpendicular to each other. A small, spherical particle is suspended at the origin. It is met by a sinusoidal wave that propagates along the x-axis. This wave oscillates perpendicular to its direction of propagation. We will call it the primary wave. It shall be a transverse wave that is confined to the xy-plane, and thus it is linearly polarized in that plane. We will assume the particle to be smaller than the wavelength of the primary wave. Once the wave reaches the particle, the latter is caused to oscillate back and forth along the y-axis, just as a cork floating in a lake moves up and down when hit by a moderately high incoming wave. The particle's small size ensures that its entire volume is affected as a unit by the incoming wave. Now we introduce an assumption of the mechanical theory of waves: in an elastic medium, all moving masses emit energy by becoming the centers of waves, and thus the particle must begin to radiate secondary waves. In principle, one may think of these waves as propagating in any direction. However, since the situation is supposed to serve as a model for the propagation of light, we shall deal with transverse waves only. All waves that would propagate along the particle's direction of oscillation, the y-axis, would be longitudinal waves, in contradiction of our assumption. However, secondary waves may propagate in any other direction, and we have not made any assumption precluding the possibility that a secondary wave may propagate in the direction of the z-axis. The secondary waves may therefore be imagined as propagating out from the particle in all directions, except the y-axis. Secondary waves within the xz-plane lack a y-component, and thus their amplitudes will be largest. Away from this plane, the intensity decreases until it reaches zero along the y-axis. There is no reason for this decline to happen abruptly, and so we may assume it to proceed gradually. The situation is symmetrical around the y-axis, and thus a donut-shaped emission pattern of secondary waves results. The part of this scattered radiation propagating in the xy-plane is depicted in Figure 7.5.

So far we have assumed that the incoming primary wave is linearly polarized in the xy-plane. Remember that Strutt wanted to show that the scattering of an incoming, *unpolarized* primary wave yields completely polarized secondary waves at right angles to the incoming wave's direction of propagation. This is what Tyndall's experiments and the observations of daylight polarization had suggested. We therefore have to contemplate the fate of an unpolarized primary wave. Such a wave can be regarded as the sum of two linearly polarized components that oscillate in phase and are perpendicular to each other. Let us assume again that the primary wave propagates along the x-axis and incites a particle to oscillate around the origin of the coordinate system. Just like the primary wave, any sec-

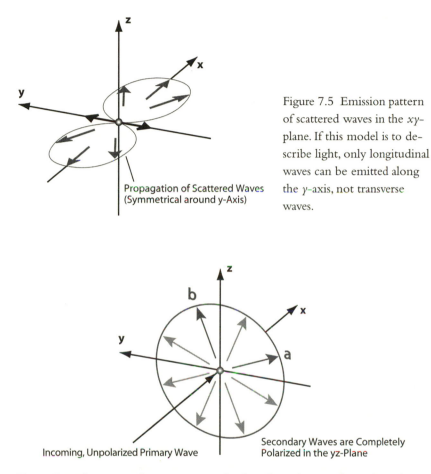

Figure 7.5 Emission pattern of scattered waves in the xy-plane. If this model is to describe light, only longitudinal waves can be emitted along the y-axis, not transverse waves.

Propagation of Scattered Waves
(Symmetrical around y-Axis)

Incoming, Unpolarized Primary Wave

Secondary Waves are Completely Polarized in the yz-Plane

Figure 7.6 The scattered waves are completely polarized at a right angle to the incoming, unpolarized transverse wave. This polarization is tangential; that is, the tangents of the scattered waves form tangents along (imaginary) circles around the scattering particle.

ondary wave may be understood as the sum of two perpendicular components. We may assume that these are parallel to the components that make up the primary waves. Let us now look at Figure 7.6. For simplicity's sake we shall only consider light scattered laterally, in the yz-plane. We call the directions of the secondary waves' components a and b. One com-

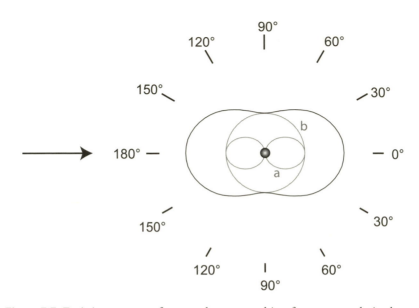

Figure 7.7 Emission pattern of scattered waves resulting from an unpolarized, transverse incoming wave. This pattern is symmetrical about the axis of propagation of the incident light ray. It is the sum of the patterns shown in figs. 7.5 (curve a) and 7.6 (curve b).

ponent of the primary wave oscillates parallel to the *a*-direction, while its other component oscillates parallel to the *b*-direction. Since the primary wave is transverse, its *a*-component cannot emit secondary waves in the *b*-direction, while its *b*-component cannot emit such waves in the *a*-direction. The light emitted in the directions *a* and *b* is, therefore, completely linearly polarized. Significantly, since we have defined *a* and *b* as arbitrary yet mutually perpendicular directions, *all* the laterally scattered light must be completely polarized. The pattern of radiation scattered by an incoming, unpolarized wave is shown in Figure 7.7. Maximum polarization occurs at right angles to the illuminating light source, just as the observations of skylight and Tyndall's experiment had shown.

This artful argumentation is not entirely Strutt's own. Rather, he borrowed much of it from Stokes, who had used it in 1852 to show that particles in the aether must oscillate at right angles to their direction of polarization (Figure 7.8). Even though the riddle of skylight polariza-

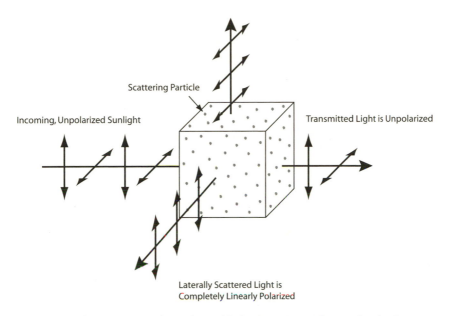

Figure 7.8 The scattering of unpolarized light waves in a volume of air leads to the pattern seen in the sky: tangential polarization with a maximum at a 90-degree angle from the sun.

tion was known at the time, Stokes himself had not realized that his clever reasoning could be applied to this phenomenon.

Having accounted for the essential properties of polarized light from the sky, Strutt goes on to consider the color of scattered light. What he aims at is an expression relating the intensity of scattered light as a function of the wavelength, assuming that the intensity of the incoming light beam is equal at all wavelengths. Strutt knows that this problem is complex, encompassing the propagation of waves in three-dimensional space as well as the particle's excitation and subsequent emission of secondary waves. Before delving into higher mathematics, he wonders whether he can guess the answer at least within an order of magnitude, a rough estimate to serve as a guide for his exact derivation. Strutt does so by contemplating the physical variables that may be relevant to the problem. Knowing that the answer should consist of a mathematical equation, and that in such an equation both sides must be of the same physical dimension, he makes an educated guess at the result. Strutt is a master of a technique that he calls the "principle of similitude," in which the physical quantities that might possibly be involved in a problem are combined in an equation in such a way that the physical dimensions of the variables, such as length, time, mass, and so on, balance each other out. Doing so involves aptly combining them as products and fractions. This technique, today called dimensional analysis, is no mere guesswork, but may help reveal the structure of a physical problem. Many famous physicists have used it to great benefit in their studies.

Strutt wants to know how the intensity of scattered light varies with the wavelength. The intensity of scattered light divided by the intensity of the incoming wave (I'/I) is a dimensionless number, but the wavelength λ has the dimension of length. Other quantities that may enter the desired equation would be the distance r of the observer from the scattering particle (dimension: length), the volume V of the scattering particle (dimension: length cubed), the speed of light (dimension: length over time), and the densities of the particles and the surrounding medium (dimension: mass over length cubed). Appendix C explains how Strutt derived his formula. Combining these quantities in a physically motivated manner, he arrives at an approximate equation for the wavelength dependence of scattered light:

$$I' \approx \frac{V^2}{r^2 \lambda^4} I$$

If the distance from the scattering particle to the observer is constant, and if the particle's volume is fixed, the intensity of scattered light varies only as a function of the wavelength of incoming light. Strutt explains the meaning of this formula as follows:

> When light is scattered by particles which are very small compared with any of the wavelengths, the ratio of the amplitudes of the vibrations of the scattered and incident light varies inversely as the square of the wavelength, and the intensity of the lights themselves as the inverse fourth power.[4]

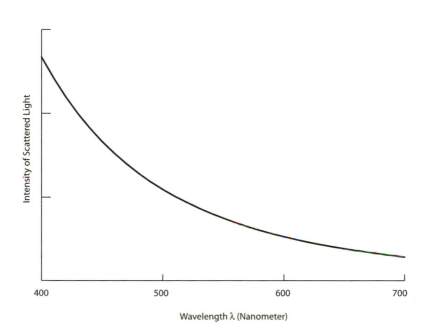

Figure 7.9 In Rayleigh's theory of light scattering, the intensity of scattered light increases dramatically with shorter wavelengths.

LIGHT SCATTERING MADE SIMPLE

The gist of Strutt's theory is that the scattering of light by small particles suspended in Earth's atmosphere can explain the color and the polarization of skylight. The intensity of scattered light strongly depends on the wavelength of incoming light, being proportional to the inverse fourth power of the wavelength. Light with short wavelengths (blue, violet) is more likely to be scattered than light with long wavelengths (orange, red). In total, the scattered light is dominated by the short wavelengths and appears blue (Figure 7.9).

Strutt regards the wavelength dependence of scattered light as a natural consequence of the relation between the scattering particles' sizes and the wavelength of incoming light. The more closely the wavelength

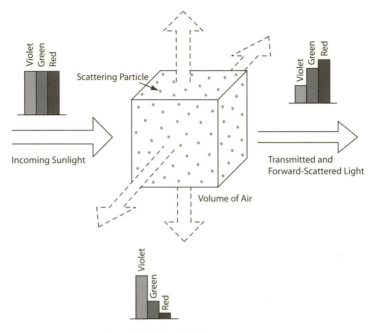

Figure 7.10 Colors of the scattered and transmitted light due to Rayleigh scattering in a volume of air.

approaches the particle's size, the more the particle will be able to affect the wave's course. In popular lectures that he delivered on a tour through the United States in 1872–73, John Tyndall came up with an intuitive picture for visualizing this process. A small pebble, if thrown into a pond, will profoundly affect the circular ripples caused by raindrops. In contrast, the pebble will scarcely affect the propagation of larger waves on the pond's surface. Much the same must hold for the influence of small scattering particles on solar rays. The short waves of blue and violet light will be affected much more than the longer waves of orange and red. The "inverse fourth power law" derived by Strutt translates this insight into the language of mathematics. Imagine a large number of independent particles that scatter light (Figure 7.10). We will not speculate on their nature yet but only assume them to be spherical, homogeneous, and much smaller than the wavelength of the illuminating light. We already know that the spectral colors are present in roughly equal proportions in sunlight, combining to make their mixture appear white. When this composite light meets the scattering particles, its spectral constituents are affected differently. Strutt's dimensional analysis approach shows that the short wavelengths in the sunlight are scattered much more than the long ones. According to the formula, the scattering intensity increases by a factor of 16 when the wavelength is halved. Thus, within the visible range of the solar spectrum, violet light with a wavelength $\lambda = 400$ nanometers is, in comparison with red light ($\lambda = 700$ nanometers), scattered (700 nm/ 400 nm)^4 = 9.4 times more intensely. Obviously, short wavelengths dominate the scattered light, and its intensity steadily increases from red to violet, with orange, yellow, blue, and green in between. The color we see in the clear sky is the additive mixture of these constituents, and as such the mixture looks blue. Sky blue.

The scattering of light waves proposed by Strutt not only accounts for the clear sky's blue color (Figure 7.11) but also explains Goethe's basic phenomenon and the colors of turbid media. The reddish color of the setting sun is one example. Close to the horizon, a ray of sunlight has to travel a particularly long distance through the atmosphere to reach an observer. Hence, the effects of scattering are intensified. Almost all of the shorter wavelengths are laterally scattered out of the ray and only the longer wavelengths remain: yellow, orange, and red. Likewise, red lights are much better suited for signaling a railroad stop, because they can be seen through a scattering medium (misty air, for example) over a longer distance than a blue light could.

Although this picture may be immediately intuitive with regard to the dominant color in scattered light, the mechanism of polarization is more difficult to grasp. Remember that without the transverse character of light waves, no polarization would result from scattering. Moreover, it is the transverse character of light that determines the pattern of scattered light. Put simply, we may conclude that of the two principal consequences of light scattering, the blue color stems from the small size of the scatterers, while the patterns of polarization and intensity are due to the transverse nature of light.

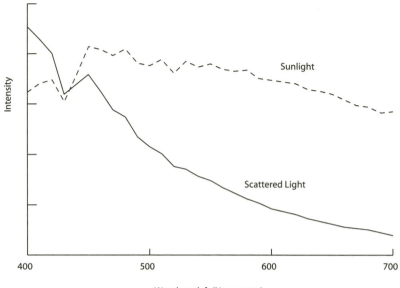

Figure 7.11 Although sunlight contains all the colors of the spectrum, they are not present in equal proportions. For this reason, the intensity of skylight does not follow the inverse fourth-power law exactly, but deviates from it slightly.

PUTTING THE THEORY TO THE TEST

We have gotten to know Strutt as a scientist who did skillful experiments on color vision. Ultimately, it was these experiments that directed his attention to the blue color of the sky. Yet even before coming up with the theory of light scattering, he sought to experimentally determine the quantitative spectral composition of blue light from the sky.[5] An elaborate setup enabled him to do so. With a mirror he reflected the blue light from the zenith onto a prism in his darkened laboratory and projected the resulting spectrum onto a white wall. To make precise measurements, he needed another light source for comparison. Sunlight served this purpose, weakened by a sheet of white paper and projected

TABLE 7.1

JOHN WILLIAM STRUTT'S MEASUREMENTS OF THE SPECTRAL INTENSITY
DISTRIBUTION OF SKYLIGHT (NOVEMBER 1870) AND THE PREDICTED
INTENSITIES ACCORDING TO HIS INVERSE FOURTH-POWER LAW OF
SCATTERING AND THE INVERSE SQUARE LAW THAT FOLLOWS FROM THE
COLOR BEING PRODUCED BY REFLECTION ACCORDING TO THE THEORIES OF
ISAAC NEWTON AND RUDOLF CLAUSIUS

Wavelength Color	Intensity Measured by Strutt	Prediction According to Strutt's Inverse Fourth Power Law	Prediction on the Assumption of Colors Being Produced by Reflection
486 nm (ice blue)	90	80	46
517 nm (sea green)	71	63	40
589 nm (yellow orange)	41	40	31
656 nm (red)	25	25	25

SOURCE: John William Strutt, "On the Light from the Sky, Its Polarization and Colour." *Philosophical Magazine*, 4th ser., 41 (1871): 114.

NOTE: Intensities are in arbitrary units. Red light serves as the reference; all intensities for this color were set to 25.

onto the same wall with another prism. Movable apertures helped him to focus on selected parts of the spectrum by controlling the amount of transmitted light. Thus, he was able to determine the relative intensities of the light from the cloudless sky and the sun for every part of the visible spectrum. Strutt's measurements (Table 7.1) match those predicted by his theory quite closely. In the four parts of the visible spectrum that he investigated, the intensity of skylight conforms almost perfectly to the inverse fourth power law that he had derived theoretically. The only deviation occurs with violet light, which seemed to be even more intense than his theory predicted. The power of these measurements becomes especially evident when compared with the predictions of the theories of Newton and Clausius, the first of which assumed the blue to be caused by interference from water droplets, the other by interference from water bubbles. Both theories predict the intensity of skylight to be inversely proportional to the *square* of wavelength. Table 7.1 indicates clearly that this yields a deficit at short wavelengths. The sky of Newton and Clausius is just not blue enough! Moreover, as we saw in Chapter 6, its predicted maximum polarization is at the wrong angle from the sun.

Strutt's 1871 article marked a turning point in the understanding of the sky's blue color, containing an elegant theory as well as very clear experimental data, whose value was soon recognized. John Tyndall congratulated Strutt in a letter dated February 3, 1871, despite feeling somewhat vexed by Strutt's criticism of his own attempts at interpreting his results. However, Strutt always emphasized that it was Tyndall's results that had led him to develop his theory. Maxwell was impressed as well. He wrote to a friend, "I think Strutt on sky blue is very good. It settles Clausius' vesicular theory."[6] Strutt must have been proud of this recognition by the greatest living physicist. In passing, he had defeated Rudolf Clausius' idea of watery bubbles floating in the atmosphere and making the sky blue by interference, a last incarnation of Newton's theory.

This recognition came when Strutt had other matters on his mind. The good news was that, in the summer of 1871 he married Evelyn Balfour, the sister of Arthur Balfour, a statesman and future British prime minister. At Cambridge, John William was a good friend of Arthur, and the former met Evelyn when he was invited by his friend to a family dinner party. Evelyn enjoyed listening to music, and when John William found out about this, he gave her his copy of Hermann von Helmholtz's book, *Lessons on the Perception of Sound* (Die Lehre von den Tonempfindungen). Her reaction is not recorded, but the gift may have given her an advance notion of what it would be like to marry a man so deeply devoted to science.

The bad news was that Strutt's health was deteriorating. He had been frail since infancy, but now he was suffering from painful attacks of rheumatic fever. His doctor advised him to find remedy in traveling to the Mediterranean. Following this advice, the couple combined a medical cure with their honeymoon by taking a trip to the Nile. One wonders what Evelyn thought about her honeymoon, considering that her husband found enough time to begin the monumental, two-volume *Theory of Sound* while traveling along the Nile in a houseboat. Published in 1877, the *Theory* was to become a standard work in the field. When the Strutts returned from Egypt in 1872, John William, devoted as ever to his studies, began to set up a physics laboratory in the horse stables of his family's country home, Terling Place. A year later, his father died. As the eldest son, John William became the third baron of Rayleigh, a title he kept from then on. It may have been strange for his family and friends to see him in this position. But he was little concerned, skipping meetings of the House of Lords in London in favour of his ambitious research projects. At the same time, he was responsible for managing the family's estates, and while most of the fields were tenured, he was relieved when, in 1876, his younger brother Edward Strutt took over this duty.

In the meantime, acclaim for Strutt's breakthrough work on light scattering grew steadily. Within a few years, his theory became accepted throughout Europe and North America. Other researchers replicated his measurements and largely agreed on the results. In the 1890s the theory of light scattering found its way into optics textbooks. Now there was a name for light scattering by small particles: Rayleigh scattering.

FLUORESCENCE

In the late nineteenth century, Strutt was not the only physicist to propose an explanation for the blue sky. Auguste de la Rive had already speculated in 1867 that skylight and its color result from the bending (refraction) of light while traversing layers of air. In 1872, an astronomer from Geneva named Étienne Alexandre Lallemand proposed that the sky's blue light stems from the fluorescence of the air, the visible light that the air's particles supposedly emit upon being illuminated by ultraviolet sunlight.[7] Observations of polarization proved decisive in evaluating both of these claims.

Mentioned in Chinese chronicles as early as 1500 B.C., fluorescence became known in Europe in 1565. The Frenchman Nicolás Monardes observed it while studying *Lignum nephriticum* ("kidney wood"), the

wood of the Mexican blue sandalwood tree (*Eysenhardtia polystacha*).
When pulverized and mixed with water it could be used to treat dis-
eases of the kidney and syphilis. Monardes discovered that this solution
emitted a strange light when illuminated by the sun. In the seventeenth
century, this phenomenon was further investigated by Robert Boyle in
London and the Jesuit scholar Athanasius Kircher in Rome. Boyle noticed that when he looked at it with the sun at his back, the solution
gave off light of a sky-blue color. When the solution was backlit, it appeared yellowish. This would seem to associate it with turbid media, but
the blue was of a purer hue and stronger intensity than the light observed in such media. Kircher realized that *Lignum nephriticum* was one
member of a larger class, and perhaps related to the glow emitted by
fireflies (which, Kircher suggested, could be used for indoor illumination). Neither Monardes nor Kircher linked this exotic phenomenon to
skylight. However, in suspecting that one kind of the particles constitut-
ing air is luminescent, Boyle may have had such a connection in mind.
By the nineteenth century, a large number of observations of strange
glowing phenomena had accumulated, from *Lignum nephriticum* and fireflies to the glow of algae in the sea and that of certain stones when
rubbed against each other.

The challenge of the day was to account for these phenomena with
the wave theory of light. George Gabriel Stokes, Strutt's optics teacher
in Cambridge, worked especially hard on this problem and published a
long paper about it. Stokes began by distinguishing between fluorescence, in which light is emitted by a substance only as long as it is illuminated by another source of light, and phosphorescence, in which a
substance continues to emit light after removal of the light source. He
confined himself to the study of fluorescence in inorganic materials,
combining experiment and theory to seek its origins. Setting out with a
solution of quinine bisulfate he soon discovered that it changed the light
in the sense that the emitted light was always of a longer wavelength
than the incident light, a finding that came to be known as Stokes' law of
fluorescence. Stokes and his contemporaries, including Lallemand, were
intrigued by the fact that fluorescent matter may emit visible light on

being illuminated by invisible ultraviolet radiation. Today this phenomenon is often used by the entertainment industry, where ultraviolet light is commonly known as "black light."

Studying the solution of quinine bisulfate more closely, Stokes realized that its light contained two components. One was apparently the result of scattering and was maximally polarized at a right angle to the illuminating light source. The other was of a "beautiful sky-blue color"—and turned out to be the fluorescent component. This is the light with which Lallemand had sought to explain the light and color of the sky.

Given that Stokes studied the polarization of light at the same time (in the early 1850s), we might expect him to have considered whether the phenomena of fluorescence and polarization are related, and indeed he did. Illuminating the solution with "polarized light in a vertical or horizontal plane as well as by common [unpolarized] light," Stokes found that the emitted, fluorescent light was never polarized.[8] He concluded that fluorescence and polarization are unrelated phenomena, and thus he had proved Lallemand wrong twenty years before the latter even proposed his hypothesis. One is not surprised, then, that Lallemand's paper was largely ignored by his contemporaries. Neither Rayleigh, nor Tyndall, nor Maxwell responded to it.

It is remarkable, though, that throughout the late nineteenth century and early twentieth century, fluorescence kept coming up as an explanation for the light and color of the sky. For instance, seventeen years after Lallemand published his paper, Walter Hartley, a chemist working at Dublin's Royal College of Science, took up the matter. Throughout the 1880s, Hartley conducted detailed laboratory investigations of the spectrum of ozone and discovered that ozone could absorb ultraviolet radiation (see Chapter 9). He noticed that ozone, once cooled into a fluid state, was of a deep blue color. To Hartley, those two common properties, namely the absorption of ultraviolet light (so typical of many fluorescent substances) and the blue color, were so striking that he could not help seeing a parallel between (fluorescent) quinine bisulfate and ozone, which, despite all efforts to identify it as such, is nonfluorescent.

FROM MAXWELL'S EQUATIONS TO DIPOLE RADIATION

By adopting a mechanical theory of light that deals with transverse waves propagating in an elastic medium, the aether, Strutt was following the mainstream of contemporary optics in 1871. In the second half of the nineteenth century, this was by far the most widely held conception of light. However, despite its popularity, this theory stood on shaky ground. Most importantly, experimental proof for the existence of the aether, which was considered necessary for conveying the light waves, was lacking. To many physicists this was not a great concern: the aether seemed so natural that they considered it just a matter of time until this medium would be discovered.

While the mechanical theory of light seemed more or less satisfactory, Maxwell, in 1864, suggested an alternative theory. It treated light as electromagnetic waves and could dispense with the aether. Parallel to his experiments on color vision, Maxwell had begun, in 1854, to think about the theory of electricity. In this branch of physics, seemingly unrelated to optics, many recent observations were still awaiting a consistent explanation. Up until the late eighteenth century, physicists had been content to deal with electrostatics, the study of electric charges at rest. A few years later it became clear that previously unanticipated effects arise when electric charges are set in motion. In 1820, the Danish physicist Hans Christian Oersted succeeded in demonstrating that electric currents produce magnetic fields. Shortly thereafter, André-Marie Ampère arrived at a mathematical description of electric ring currents. In 1831, the year of Maxwell's birth, Michael Faraday discovered the law of electromagnetic induction. These important contributions revealed that electric and magnetic fields are closely related.

Electromagnetism had been discovered, but a theory linking electricity and magnetism was lacking. Maxwell succeeded in solving this difficult problem by proposing a system of mathematical equations which predicts that a varying electric field will produce a magnetic field, and a varying magnetic field will produce an electric field. Since the electric and magnetic fields generate each other directly, no medium translating one field into another is necessary. Nevertheless, Maxwell was not to give up his belief in the existence of the aether.

Before long, Maxwell discovered an intriguing property of his equations. In the absence of electric currents and charges, they assume the mathematical form of wave equations, with transverse electromagnetic waves as their solutions. By plugging in known electric and magnetic quantities, he was able to predict the speed of these hypothetical waves: 300,000 kilometers per second. This is exactly the speed of light. This surprising corroboration strongly suggested to Maxwell that light itself could be an electromagnetic wave. He was on his way to a momentous discovery. If he was correct in this conjecture, electricity, magnetism and optics, previously thought to be unrelated phenomena, must have a common basis. Full of excitement, he wrote to his cousin on January 5, 1865, "I have also a paper afloat, containing an electromagnetic theory of light, which, till I am convinced to the contrary, I hold to be great guns."[9]

In 1887, Maxwell was proven right when Heinrich Hertz succeeded in producing electromagnetic waves and showing them to behave just like light waves as far as their reflection, refraction, and polarization were concerned. Maxwell's equations were proven beyond the shadow of a doubt. They are among the greatest achievements in the history of physics.

The theory of light scattering devised by John William Strutt, the third Lord Rayleigh, can be translated into the framework of Maxwell's theory, and it was Rayleigh himself who took on the task in 1881. In order to do so, he had to replace his mechanical model with electromagnetic descriptions of both the wave and the particle. In Maxwell's theory, light is assumed to consist of transverse electric and magnetic fields, periodically oscillating at right angles to each other. Figure 7.12 shows how this can be envisioned. We can assume the scattering particle to contain both positive and negative electric charges. When affected by the incoming wave's electric field, these charges are displaced within the particle. As shown in Figure 7.13, the particle is electrically polarized and has become an electric dipole. Since its charge distribution is periodically reversed due to the incoming wave, the particle will act as an electric oscillator, re-radiating the received energy (Figure 7.14).

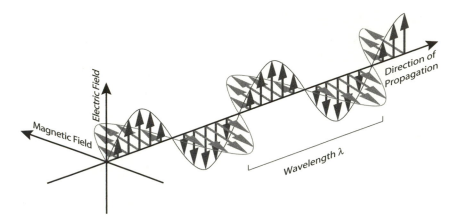

Figure 7.12 Propagation of light as an electromagnetic transverse wave according to Maxwell's theory. The electrical and magnetic components are perpendicular to each other.

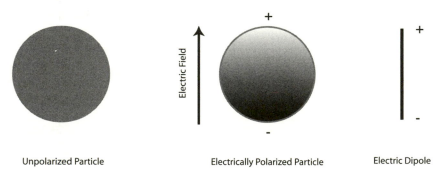

Figure 7.13 In an electrically unpolarized particle, positive and negative charges are distributed uniformly. An electric field displaces these charges with respect to each other; the particle becomes polarized as an electric dipole.

At first the electric field lines describe arcs from the dipole's positive to its negative charge. Half a period later, its charge distribution is reversed. Now the field lines separate from the dipole, close in on themselves, and propagate away from the dipole at the speed of light. The reversal of the charge distribution produces a new set of field lines having the same

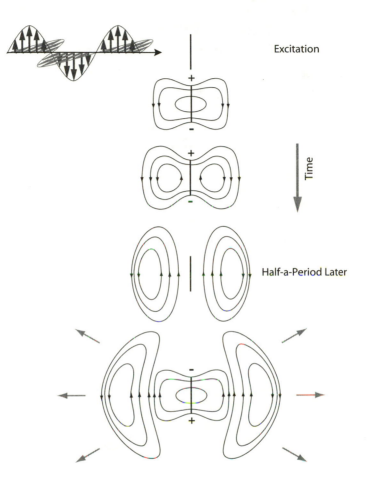

Excitation

Time

Half-a-Period Later

Figure 7.14 When a small, polarizable particle is acted on by an electromagnetic wave, it becomes a dipole whose charge distribution reverses itself periodically. The dipole sends out bundles of electric and magnetic field lines, each of which separates and is emitted after half a period of the incoming wave; the particle has become a Hertzian dipole. The field lines that are emitted make up the scattered waves. The electric field lines are shown here. The magnetic field lines (not shown) form circles around the axis of the dipole. All field lines are symmetrical about the dipole's axis, along which no waves are emitted.

form but opposite directions. Accelerated electric charges always produce a magnetic field, and in our case its field lines may be represented as circles around the dipole's axis. These are coupled with the electric field lines, and all together produce transverse electromagnetic waves propagating away from the dipole at the speed of light. This is the scattered light.

Stimulated by the incoming electromagnetic wave, the particle will continue to act as an electric oscillator until the received energy has been re-emitted completely. Therefore, scattering does not imply absorption, nor is the energy lost. Rather, it causes the light to be redistributed according to the pattern shown in Figure 7.7. In this theory only small particles are polarized uniformly. Larger particles will be polarized unevenly by an incoming light wave and do not act as electric dipoles.

Maxwell did not live to see Rayleigh integrate light scattering into his theory of electromagnetism. In 1879, the famous Scotsman had died prematurely at the age of 48. Rayleigh was asked to succeed him as Cavendish Professor of Experimental Physics at Cambridge. Initially he hesitated, preferring to go on with his routine at Terling Place. However, in the late 1870s a severe agricultural depression hit England, and the revenues from the farm were in decline. Rayleigh then accepted the post, and held it until 1884. During his five years as the head of the Cavendish laboratory, Rayleigh reformed the curriculum, introducing compulsory classes in laboratory physics, and launched a program to standardize electrical units.

A GUESS ON THE SCATTERING PARTICLES

While the theory of light scattering by small particles was an immediate success among Strutt's colleagues, the nature of the scattering particles in Earth's atmosphere remained elusive in 1871. Color and polarization, the two observable properties of light scattering considered by Strutt, are the same for all particles, provided they are spherical, much smaller than the wavelength of the illuminating light, and sufficiently numerous. These properties of skylight do not reveal much about the nature of the scatterers. Strutt guessed that small particles of "common salt" floating in the air

TABLE 7.2
CHEMICAL COMPOSITION OF THE EARTH'S "DRY" ATMOSPHERE

Gas	Chemical Symbol	Percent by Volume
Nitrogen	N_2	78.08
Oxygen	O_2	20.95
Argon	Ar	0.93
Carbon dioxide	CO_2	0.0370 (in 2001)
Noble gases	Ne, He, Kr, Xe	0.002

SOURCES: Richard P. Wayne, *Chemistry of Atmospheres*, 3rd ed. (Oxford: Oxford University Press, 2000), p. 2; John Houghton, *Global Warming: The Complete Briefing*, 3rd ed. (Cambridge: Cambridge University Press, 2004), 16.

NOTE: The variable content in water vapor is not included here.

may account for the scattered light in the atmosphere, knowing that he could not prove this with the evidence available to him.

One may liken the challenge that Strutt faced in 1871 to the task of inferring the shape of an unknown dragon from its footprints.[10] With regard to the atmosphere, the situation is further complicated by the large number of scatterers required to create diffuse skylight. These amount to a flock of many small dragons—an enormously difficult task that Strutt deemed impossible. It seemed much easier to identify the tracks of a dragon with an already known shape, so it is not surprising that Strutt tried this approach.

Relating the footprints' shape (the observed light of the sky) to the flock of dragons (the scattering particles) requires some knowledge of the composition of atmospheric air. Since the days of Humboldt and Gay-Lussac's work on the chemical composition of air, little improvement had been made in the gas fractions determined by these two scholars—this was to happen only with the discovery of Argon in 1894. But it had become clear that water vapor is subject to considerable day-to-day changes and may vary between small traces (in desert air) and 7 percent (in humid climates), amounting to a teaspoon of water in a cubic meter of air.

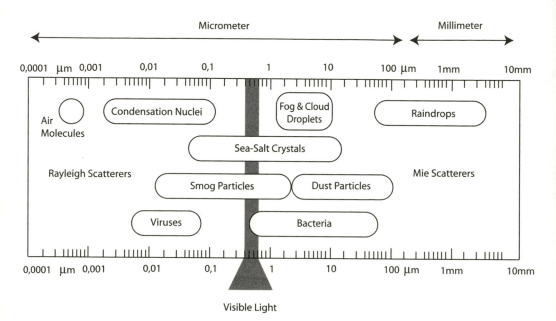

Figure 7.15 Size distribution of typical scattering particles in the atmosphere. Particles that are considerably smaller than the wavelength of the light act as Rayleigh scattering particles. Larger particles scatter light in accordance with Mie's theory. Modified after V. J. Schaefer and John A. Day, *Peterson's Field Guides: Atmosphere* (New York: Houghton Mifflin, 1981), 4.

Researchers understood that pure air, as listed in Table 7.2, does not exist anywhere. Rather, it is always polluted by a number of so-called aerosol particles which occur in a wide range of sizes (Figure 7.15). Their concentrations vary depending on place and time. Dust whirled up by the wind, salt freed from oceanic surface water, smoke from fires, volcanic dust, as well as pollen and even bacteria and viruses are examples of natural aerosols. Every few years, dust storms in the Sahara can stir up dust that is subsequently carried on the wind to Central Europe and across the Atlantic Ocean. Aerosol particles are found mostly in the lower layers of the atmosphere and fall back to Earth's surface within a short time. How long they are suspended in the atmosphere depends on their size, mass, and elevation above the ground.

Considering the size distribution of aerosol particles, we can understand Strutt's caution in suggesting "common salt" as possible scatterers. Although sea salt crystals are indeed small enough to scatter visible light, many other aerosols can do so as well, such as cloud condensation nuclei, viruses, and tobacco smoke.

MULTIPLE SCATTERING, MIE SCATTERING, AND WHITE CLOUDS

Given the advances in colorimetry made in the nineteenth century, it was possible to examine skylight more closely—and to compare measurements with the predictions made by Strutt's theory. Maxwell's experiments with spinning color tops had reinforced the idea that a color may be characterized by its hue, saturation, and tone (or luminosity). The term hue usually denotes a color with regard to the taxonomy of color names, or, for physicists, its wavelength. Measurements of typical blue skylight showed that it most closely resembles an ultramarine blue with a dominant wavelength of 475 nanometers. Saturation denotes the percentage of pure hue in a mixture of colors. Strutt's theory predicts that pure blue skylight should have a saturation of 42 percent. This relatively small value is due mostly to the wavelength dependence of scattered light, as shown in Figure 7.9. However, measurements of even the clearest blue showed that its saturation is always much less than 42 percent. The tone (or luminosity) of skylight was not readily evident from the first measurements but was to become the subject of later investigations that are described in the next chapter.

The discrepancy between the actual and the predicted saturations of skylight hinted at a deficiency in Strutt's theory. The finding that polarization at right angles to the sun is not complete, as single Rayleigh scattering predicted, was another indication. Observations revealed that it is at most 70 percent.

Strutt himself was aware of the problems posed by these observations, and he had a solution in mind. For simplicity he had assumed that each ray that enters our eyes was scattered by just one scatterer on its entire passage through Earth's atmosphere. Given the large number of scatter-

ing particles in the atmosphere, it seems inevitable that a light ray will be
scattered not only once, but several times. In other words, the atmosphere may not only be illuminated by the sun, but it also illuminates itself. This process is called multiple scattering. Even though each individual instance of scattering may accord to his theory, Strutt was aware in
1871 that the total effect of multiple scattering must alter the light of the
sky with respect to a prediction based on single scattering. Yet he also
knew that it was extremely difficult to calculate its properties. Only with
the advent of modern electronic computers have realistic computations
of multiple Rayleigh scattering become feasible.

Complicated as an exact characterization of multiple scattering may
be, a simple experiment proposed by the meteorologist Craig Bohren
gives us an idea of its impact. Shine white light onto the side of a glass
filled with clear water in front of a dark background. If you drip a little
milk into the water, the suspension appears bluish, much as in Brücke's
experiment (Chapter 6). The difference between the two is that with
the milk, it is the small fat droplets that scatter the light, whereas in
Brücke's experiment it was the particles of pistachio resin. But if more
milk is added, the suspension appears increasingly white, like the familiar appearance of ordinary milk (Color Plate 20). The fat droplets have
not changed their properties; they have merely increased in number—
and multiple scattering has become more significant. Without multiple
scattering, milk would look blue.

Similarly, a larger number of scatterers would make the clear sky look
paler. This can be seen in clouds and down near the horizon, because
there is a larger number of air particles in this direction than toward the
zenith. That is why the sky over the horizon is lighter in color than at
the zenith. But why is the blue less saturated toward the horizon? To understand this, we cannot simply extend the conclusions drawn from the
milk experiment.

Imagine an atmosphere that consists solely of Rayleigh scattering
particles. The light that comes to us from a particular direction is affected
by many scatterers in this direction. If we look high up in the sky, around
the zenith, there are relatively few scattering particles between us and the
upper edge of the atmosphere (Figure 7.16). From this direction, it is

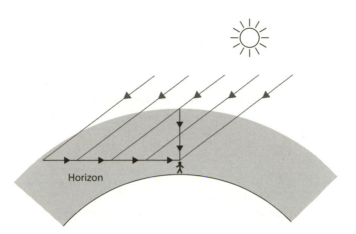

Figure 7.16 Skylight just above the horizon results from scattering at a large number of particles. This leads to a pronounced brightening toward the horizon. Moreover, the additive mixture of the different components of the scattered light dilutes the blue color at the horizon. Conversely, the blue color is especially vivid at the zenith, because there are relatively few particles in this direction.

mostly singly scattered light with short wavelengths that reaches our eyes. Owing to the thickness of the atmosphere and the curvature of Earth, the path of a light ray reaching our eye from near the horizon is thirty-five times as long as it is near the zenith. Accordingly, the light coming from this direction is the sum of scattered light from thirty-five times as many scattering particles! The whitening at the horizon is the result of this difference. A certain fraction of the scattering particles located near the horizon will send singly scattered light with short wavelengths to us. But scattered light with long wavelengths will also reach us from this direction, because it can pass through greater depths of air without being rescattered. The light near the horizon, then, is the additive mixture of the scattered light from the many scattering particles in this direction, and since it contains all wavelengths in similar proportions, it ought to appear white. This is exactly what we see.

Multiple scattering is involved not only in the whitening of the sky near the horizon, but also in making white clouds and air pollution look

innocently white. In these instances it is joined by another process called
Mie scattering. It is often impossible to distinguish between these two
effects with the naked eye. A number of different particles in the atmosphere are too big for Rayleigh scattering to apply. Fog and cloud
droplets, dust, and large aerosols are examples. These particles do have
the potential to scatter light as well, but already Rayleigh realized that
they must do so in ways different from those of smaller particles. The
reason is simply that these particles are too big to be electrically polarized as a unit by an incoming light wave, and this prevents them from
becoming oscillating dipoles. It was the German physicist Gustav Mie
who, in 1908, presented a theory of light scattering which extends
Rayleigh's theory to larger scattering particles. When the scatterers are
larger than about one-tenth of a wavelength, Mie scattering has to be
taken into account. Its mathematics is quite involved, so I shall mention
only its essential features.

The intensity of Mie-scattered light depends strongly on the direction, with forward scattering dominating for larger particle sizes. The
polarization of this light is less pronounced than in Rayleigh scattering,
but depends more on direction. An important contrast to Rayleigh scattering is that in Mie scattering, differences in the wavelength do not affect the intensity of the scattered light much. In many instances, Mie
scattering of incident white light produces white scattered light as well.
Condensed water vapor and cloud droplets are typical Mie scatterers,
and thus humid air suppresses the saturation of the sky's blue. Most
clouds are white, since they scatter sunlight without changing its spectral
composition. Near industrial plants the sky often looks milky white. In
both cases multiple scattering predominates, but Mie scattering is involved as well.

Molecular Reality

The year is 1908. Jean Perrin is performing an intricate experiment in his laboratory in the rue Cujas in Paris. Perrin (Figure 8.1), a physical chemist at Sorbonne University, is studying solutions of tiny granules of gamboge, yellow vegetable latex, in water. He places the granules in a film of water 0.12 millimeters thick, about the diameter of a human hair, wondering how the particles will be distributed after settling for some time. The gamboge granules are exceedingly small, measuring one-thousandth the diameter of a human hair. Several hours after preparing the solution, Perrin observes the vertical distribution of the granules through his microscope. He realizes that the particles are neither distributed uniformly throughout the volume of the water, nor have they all settled to the ground. Rather, they are more densely concentrated toward the bottom, growing progressively sparser toward the top (Figure 8.2). Perrin counts the number of gamboge particles in his microscope's field of view at different heights in the solution. The resulting set of numbers shows that the height density of the particles follows a regular geometric progression: the density at the top is one-half that at the bottom. A similar law is known from Earth's atmosphere, whose density is halved with an ascent of 5,600 meters. The distribution of the granules in the water appears to be a model for the vertical structure of Earth's atmosphere.

On close inspection with the microscope, Perrin sees that the individual gamboge granules are not standing still. Far from it: each of them seems to jiggle around randomly, a motion that would not come to rest even after the solution has been left untouched for days or even weeks (Figure 8.3). Perrin investigates the statistical phenomenon of height distribution in an attempt to comprehend the bizarre microscopic motion that has baffled scientists since the 1820s.

Figure 8.1 Jean Perrin in a photograph from the late 1920s. Courtesy of
Archiv für Kunst und Geschichte, Berlin.

Studying plant reproduction in 1827, the Scottish botanist Robert
Brown viewed a suspension of spherical pollen granules of the evening
primrose (*Clarkia pulchella*) in water under his microscope. At large
magnification, he noticed that these tiny particles moved in a constant,
apparently random zigzag manner. Just a year earlier the Frenchman

Figure 8.2 Vertical distribution of gamboge (yellow vegetable latex) particles in a water solution, observed by Jean Perrin several days after its preparation. The number of particles decreases exponentially with increasing height, much like the vertical number density of particles in Earth's atmosphere. Reproduced from Jean Perrin, *Les Atomes* (Paris: Librairie Félix Alcan, 1913), 144.

Adolphe Brongniart had seen this motion, and attributed it to the pollen being alive. The fact that the motion seemed to become more vigorous when Brongniart raised the temperature of the suspension surely indicated an increase in its vitality, or so he believed. He had ruled out the possibility that the motion was caused by currents in the fluid, its surface capillarity, or evaporation.

Brown was able to confirm some of Brongniart's conclusions, but he went on to challenge the Frenchman's idea that the motion was due to the pollen granules being alive. While Brongniart had confined his studies to fresh pollen, Brown extracted additional specimens from herbaria that were between twenty and one hundred years old. Though hardly still alive, they exhibited the same kind of random zigzag motion in

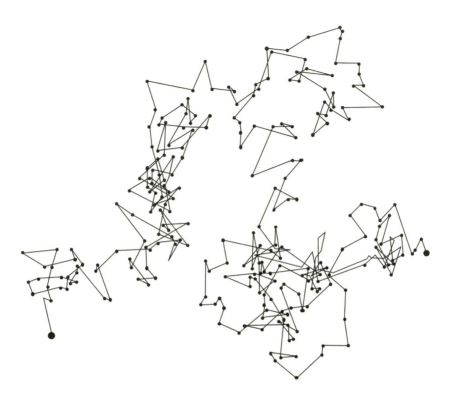

Figure 8.3 Random motion of a gamboge particle in a water solution as observed by Jean Perrin. Reproduced from Jean Perrin, *Les Atomes* (Paris: Librairie Félix Alcan 1913), 166.

water. Brown went on to investigate a variety of soils, metals, rocks, and even a fragment of Egypt's famous Sphinx near Cairo. These substances were certainly not alive, but they nonetheless all exhibited the very same zigzag motion. No wonder the discussion of Brownian motion, as the phenomenon was to be called, left biology and entered physics and metaphysics.

Even though it was studied off and on throughout the nineteenth century, profound interest in Brownian motion resumed only in 1895, when Louis-Georges Gouy realized that it provided a "natural laboratory" to study the microscopic nature of matter: is it continuous or does it consist of chemically indivisible particles, the atoms? It was Gouy's work that inspired Perrin to devise sophisticated new techniques for studying the phenomenon in the hope of proving that atoms and molecules exist. He repeated his precise determination of the granules' height distribution thousands of times, and amassed an enormous bulk of statistical data.

By the end of the nineteenth century most scientists had become convinced that atoms and molecules were real, but in crucial respects this acceptance was based on intuition rather than proof. Ernst Mach and Wilhelm Ostwald were perhaps the two most notorious critics of the prevailing view, the latter insisting that matter does not consist of atoms but energy, and that energy is divisible in any arbitrary proportion. Robert Brown had already discussed his observations in terms of the motion of molecules, a concept introduced in 1808 by the Italian chemist Amedeo Avogadro. The latter had studied chemical reactions of gases, including air, and was the first to come up with the notion of air molecules. Three years previously, Humboldt and Gay-Lussac had determined the chemical composition of air with a precision that was to be improved only late in the nineteenth century, most notably the discovery of the element argon by Rayleigh and William Ramsey; it takes up nearly 1 percent of the air's volume. This was a breakthrough for which Rayleigh and Ramsey were awarded the 1904 Nobel prizes in physics and chemistry.

In the meantime, the nature of air at the microscopic level remained a mystery. In 1871, when John William Strutt, the soon-to-be third Lord Rayleigh, had suggested that "common salt" could be the scatterer of light in the atmosphere, he evaded the question of what the air was. Although air is predominantly made up of gases (see Table 7.1), Strutt does not mention them as possible scatterers, invoking instead an observation made by Tyndall, who despite repeated attempts had been unable to detect any light scattered laterally by purified air. At the same time, statistical mechanics was providing clues to the reality of air molecules. This theory, which Rudolf Clausius and James Clerk Maxwell had been working on since the 1850s, explains the macroscopic properties of gases, such as temperature, pressure, and density, as the statistical effects of collisions between their many tiny molecules, somewhat like balls colliding on a pool table. Based on certain assumptions within the theory, it seemed possible to infer something about the size, number, and speed of gas molecules from macroscopic observations. By measuring the change in volume of vaporized fluids, the Austrian physicist Johann Joseph Loschmidt thus managed, in 1865, to estimate the diameter of air molecules to about a nanometer—a millionth of a millimeter. That is one five-hundredth of the wavelength of blue light. Even though this figure was still uncertain, it suggested that air molecules are considerably smaller than the wavelengths of visible light.

Two years after Rayleigh had presented his theory of light scattering, Maxwell became interested in the size of air molecules and arrived at the even smaller figure of 0.5 nanometers, one-half of Loschmidt's estimate. This at least served to confirm that molecules, if they existed, must be minuscule compared with the wavelength of visible light. Maxwell searched for other ways to infer their size from macroscopic observations. If independent methods should indicate the same value, he thought, then this would provide strong support for the validity of statistical mechanics and the existence of molecules. Eventually, Maxwell remembered Rayleigh's theory of light scattering. Nitrogen and oxygen take up 99 percent of the volume of air. Wouldn't their molecules be likely candidates for the scattering particles? Reason enough to ask Rayleigh whether optical observations of the sky could provide new insights into its molecules. In a letter dated August 28, 1873, Maxwell wrote:

> Suppose that there are N spheres of density ρ and diameter s in unit of volume of the medium. Find the index of refraction of the compound medium and the coefficient of extinction of light passing through it. The object of enquiry is, of course, to obtain data about the size of the molecules of air. Perhaps it may lead also to data involving the density of the aether.[1]

Maxwell knew that, up until then, Rayleigh had only looked at the color and polarization of skylight. So far, he had hardly touched on the attenuation of light over its path through the atmosphere and the influence of scattering particles on the refractive index of air. It was conceivable that new insights lay concealed there.

EXTINCTION

Maxwell's question was well posed. When a ray of light is attenuated due to its passage through an optical medium, physicists and astronomers refer to it as extinction (from the Latin *extinguere*, to extinguish). Scattering does indeed cause extinction, for it diverts a part of an incoming light ray from its original course and thus reduces the intensity of light remaining in the ray (Figure 8.4). The more scattering particles are present, the more pronounced is the extinction. An everyday example is the way the sun appears less bright when near the horizon. Similarly, the moon, the planets, and the stars appear dimmer when they are low in the sky.

In addition to weakening a light ray by diverting its intensity in all directions, the transmission of a light ray through a scattering medium causes the ray to become reddened. A well-known example of this effect is a red sunset. This is due to the blue rays being scattered out of the direct sunlight during its long passage through the atmosphere. The long wavelengths of yellow, orange, and red light are less affected, being able to slip past the many scattering particles and eventually reach the observer. Rayleigh scattering inevitably changes the color of the transmitted light.

At the time of his premature death in 1879, Maxwell had not received an answer from Rayleigh. We know today that Rayleigh kept pondering the problem off and on but was not ready to publish his material until 1899, more than a quarter of a century after receiving Maxwell's letter. Maxwell had been less interested in the color change of the transmitted light than in its attenuation due to its passage through the atmosphere. Starting out from the theory of light scattering, Rayleigh derived an exponential law that describes the weakening of the intensity of a light ray. This law relates the attenuation of light to the refractive index of air, the

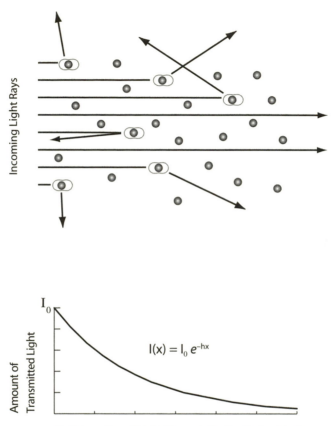

Figure 8.4 When it passes through a medium that contains small particles, a light
ray is attenuated owing to Rayleigh scattering (extinction). The more scattering particles there are, the stronger the extinction. Rayleigh's law of light scattering makes
it possible to calculate the number of scattering particles per volume of air from the
degree of extinction. See Appendix D for a derivation of the exponential law.

number of scattering particles per unit volume, and the wavelength (see
Figure 8.4). Given the index of refraction and the wavelength, the scatterers' number density can be determined from extinction. Conversely,
extinction can be determined from number densities as determined by

other means. Rayleigh chose the second route, using Maxwell's estimate of about 1.9×10^{19} molecules per cubic centimeter at sea level. It
is impossible for us to imagine such a large number. In words it is 19
quintillion or 19 million million million.

Assuming the air does indeed contain this staggering number of scatterers per cubic centimeter, Rayleigh calculated that scattering should
weaken a light ray to 37 percent of its original intensity when it passes
through 83 kilometers of atmospheric air. Was this a realistic estimate?

It probably was, or so Rayleigh thought, after he heard from British
colonial officers stationed at the Indian hill station Darjeeling that it was
possible to see the snow-capped summit of Mount Everest 160 kilometers away. According to his own calculations, the intensity of the light reflected off the snow fields should have been diminished to 15 percent of
its original value. Given the capability of our eyes to discern white surfaces even when the contrast is reduced to about 5 percent, Mount
Everest should be visible at distances of up to 250 kilometers. Thus, the
reports from India were in reasonable agreement with Rayleigh's expectations, but they did not place a significant constraint on the number
density of the air molecules as determined by Maxwell.

A better way to constrain the number density of air molecules was to
use the observations that Pierre Bouguer had collected on November 23,
1725, ten years prior to his expedition with his compatriot Charles-
Marie de la Condamine to the slopes of Chimborazo (see Chapter 5).
Bouguer was a child prodigy who had succeeded his father as Royal Professor of Hydrology at the age of 15. On that November day in 1725, he
was to study light attenuation in the atmosphere by making a set of
naked-eye observations of the moon. Figure 8.5 shows how he proceeded. Ignoring the curvature of the Earth and treating the atmosphere
as a thin plate, Bouguer realized that he needed to observe the brightness of the moon at two different positions in the sky, when the moon's
light rays traversed atmospheric layers of different thicknesses. If the
moon were at the zenith (which it never is in Brittany, northwestern
France, where Bouguer did his observations; but let us assume so for the
sake of simplicity), its light would traverse the entire thickness of the atmosphere, no more and no less. With increasing angular distance from the

zenith, the length of the light's path through the atmosphere is increased, and more of the moonlight is attenuated along the way (Figure 8.5). Thus, with the moon at an angular distance of 60 degrees from the zenith, its light rays travel through twice the amount of air as when the moon is at the zenith, and thus they are attenuated proportionally more. In principle, measurements at two different positions should suffice to determine the atmosphere's capability to attenuate light from outer space. To measure the intensity of moonlight, Bouguer compared it with a candle, making them of the same apparent brightness by placing the candle at varying distances. He found that atmospheric extinction at the zenith amounts to about 20 percent.[2]

In 1899, this number seemed credible, having since been confirmed by other researchers. Plugging it into Rayleigh's equations, we find that a light ray will be attenuated to 11 percent of its original intensity after passing through 83 kilometers of the air at sea level. This is but one-third of what Rayleigh had calculated from Maxwell's number density of air molecules at sea level. However, part of the relatively large extinction measured by Bouguer must have been due to "foreign matter," such as water vapor or aerosol particles, that is always present in the atmosphere but is not accounted for by Maxwell's value. Rayleigh inferred that the observed extinction, as well as the light from a cloudless sky, can be reasonably well understood solely in terms of the light scattered by air molecules:

> We may conclude that the light scattered from the molecules
> would suffice to give us a blue sky, not so very greatly darker than
> that actually enjoyed.[3]

So it was plausible that the scattering of light by pure air alone, free of "foreign" aerosol particles and water droplets, could explain the brightness, color, and polarization of skylight: the sky is blue because air molecules scatter the sunlight! Indeed, their small sizes and large numbers make air molecules the ideal Rayleigh scatterers. In retrospect we can see that it was circumstantial to implicate dust (al-Kindi), warm moisture (Leonardo), water and ice droplets (Newton), water bubbles (Clausius), floating particles (Tyndall), or salt particles (Rayleigh in 1871) in gener-

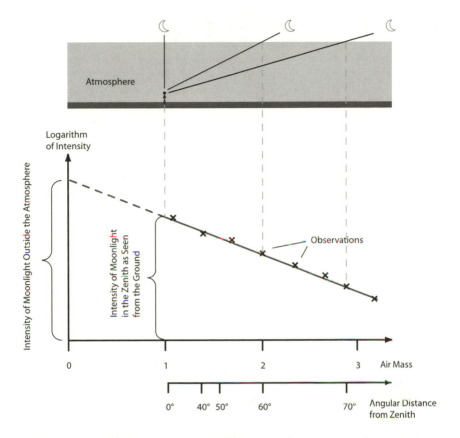

Figure 8.5 Pierre Bouguer determined the degree of extinction in the atmosphere from the decrease in the apparent intensity of light from the setting moon. This decrease depends on the quantity of air the light passes through on its way through the atmosphere. The air mass, for which the basic unit is the amount of air in the direction of the zenith, increases by a factor of 35 at the horizon. Assuming that this trend continues down to an air mass of zero, Bouguer calculated the extinction of starlight in the atmosphere.

ating the blue skylight. After all, only the air molecules are present everywhere on Earth, at all times and in invariable proportions.

Even after 1899, Rayleigh did not believe that scattering by air molecules could be detected in a laboratory. It was not until 1915 that Jean Cabannes, a Parisian physicist, succeeded in observing light scattering by pure air in the laboratory. Three years later, in 1918, Rayleigh's own son Robert John Strutt demonstrated that this light has the characteristic properties of Rayleigh scattering with regard to its spectral distribution as well as its polarization (Figure 8.6). Half a century after Tyndall claimed to have brought the blue sky into the laboratory, his assertion was borne out.

AVOGADRO'S NUMBER

Thus far, Rayleigh could only argue that molecular scattering is the source of skylight. But was it possible to infer the molecular composition of matter from observations of the sky? To answer this question, we must trace the argument step by step and consider how the molecular hypothesis evolved in the nineteenth century.

In a lecture presented to the Paris Academy of Sciences in 1808, Joseph Gay-Lussac reported on his experiments with chemical reactions between various gases. He had observed that the volumes of the gases consumed and produced occur in integer quantities, when measured at a contant pressure and temperature. In the same year, the Englishman John Dalton claimed that all elements are composed of imperceptible atoms. These atoms, he suspected, are responsible for the specific element's chemical behaviour. Chemical reactions, Dalton argued, result from the combination of atoms of two or more elements. A molecule was thought to be a particle in which two or more atoms are combined.

Avogadro used these hypotheses in 1811 to explain Gay-Lussac's experimental results. He did so by postulating that at a specific temperature and pressure, identical volumes of any given gases contain the same number of molecules. Later known as Avogadro's law, this regularity not only accounts for Gay-Lussac's results, but made it possible to determine the relative masses of the atoms. In general, equal volumes of any two chemically different gases will differ in mass. Assuming the composition of the molecules is known, the masses of the gas volumes can be translated into a scale of relative atomic weights. For instance, if we take the mass of hydrogen—the lightest chemical element—as the unit, the relative

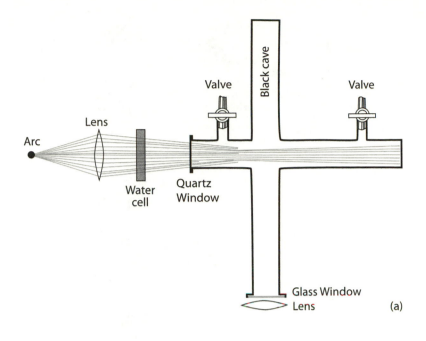

Figure 8.6 In 1918 Robert John Strutt was able to prove that pure air without foreign objects in it scatters light according to the theory derived by his father. He did
so by filling a container with carefully filtered air, then illuminating it with an intense
beam of light. He verified the laterally scattered light photographically; because of
the low intensity of the scattered light, he had to expose the film for up to twenty
hours. Since even the blackest coat of paint reflects some light, and therefore was not
a dark enough background, he built a dark shaft into his apparatus to serve this purpose (a). Just like looking into a tunnel, this shaft provided an extraordinarily dark
background. Figure b shows a photograph of the scattered light through an ultraviolet filter: it is clearly visible as a horizontal bar. The circle of light surrounding it is
caused by light reflected within the apparatus. When viewed through a yellow filter,
however, the scattered light cannot be seen (c). The scattered ray must therefore consist of predominantly short wavelengths. Figure d shows the ray without a color filter,
but with a polarizing filter that is transparent to vertically polarized light; here the
light ray is clearly visible. But when rotated 90 degrees, the filter does not let the
scattered light through (e). The scattered light must therefore be tangentially polarized. The color and polarization of the scattered light thus conform to the properties
of Rayleigh scattering. Reproduced from Robert John Strutt, "Scattering of Light by
Dust-free Air, with Artificial Reproduction of the Blue Sky—Preliminary Note,"
Proceeding of the Royal Society of London, ser. A, 94 (1918): 455 and Plate 4.

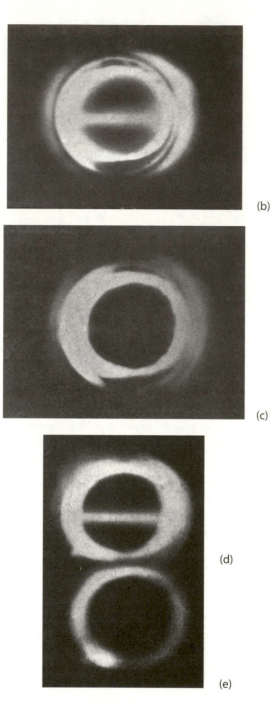

(b)

(c)

(d)

(e)

Figure 8.6 (*continued*)

atomic masses will be 12 for carbon, 14 for nitrogen, and 16 for oxygen. Therefore, one gram of hydrogen must contain as many atoms as 12 grams of carbon or 16 grams of oxygen. The number of atoms contained in the "gram atom" (also called a mole) appears to be an elementary property of nature—it relates the microcosm of atoms to the macrocosm of balances and weights. If atoms exist, this so-called Avogadro's number must be one of the fundamental constants determining the structure of matter in the universe. As such it ranks as high as the speed of light, the gravitational constant, or the electron's charge.

Now, Avogadro's number can be determined by measuring atmospheric extinction; it is simply a matter of reversing the calculation done above. This is just what Rayleigh did, only to conclude that the number density of air molecules determined by Maxwell could adequately account for the extinction observed by Bouguer and the British troops in the Himalayas. Rayleigh obtained a lower limit for the number density of air molecules at sea level that agreed reasonably well with that determined by Maxwell: at least 7×10^{18} per cubic centimeter. Rayleigh was unaware that the Danish physicist Ludwig Valentin Lorenz had already used light scattering toward the same end in 1890, arriving at a number density of 1.63×10^{19} molecules per cubic centimeter. In retrospect, this value is slightly better than what Maxwell had found. But, published in Danish in the *Publications of the Royal Danish Scientific Society*,[4] Lorenz's results went unnoticed in England.

We may assume that air is an ideal gas, that is, it is defined by the fact that its molecules are small and randomly distributed through space, that these molecules are in constant motion, colliding with each other elastically, and do not attract each other by other forces. Given this condition, Avogadro's number can be computed from the number density of air molecules per cubic centimeter at sea level by simply multiplying the latter number by 22,414 cubic centimeters per gram atom. This factor is the so-called molar volume, the volume of one gram atom of gas molecules at sea-level pressure and the freezing point of water ($0°C$). Thus, the number density of air molecules as determined by Ludwig Valentin Lorenz in 1890 implies an Avogadro's number of 3.65×10^{23} atoms per gram atom. This number was well in line with other contemporary observations. By

1900, physicists had found a variety of different experimental techniques to measure the value of Avogadro's constant, yielding numbers between 10^{22} and 10^{24} atoms per gram atom. Unlike Maxwell, Loschmidt, and Clausius, not everybody was convinced that these numbers converged toward one and the same value, as the hypothesis of molecular reality demanded.

BROWNIAN MOTION EXPLAINED

When Jean Perrin began his studies on Brownian motion, he was a thirty-eight-year-old reader in physical chemistry at Sorbonne University. After growing up in modest circumstances in northern France, he was admitted to the prestigious École Normale Supérieure in Paris and earned his doctorate there with research on cathode rays and X-rays, which had recently been discovered by Wilhelm Conrad Röntgen. This early work established Perrin as a proponent of the molecular hypothesis and earned him a prize from the Royal Society. Perrin then accepted a position at the Sorbonne, where he was to develop a course in physical chemistry. Perhaps it was this project that led him to contemplate a study of Brownian motion in 1906. Recent work, facilitated by the invention of the ultramicroscope—an instrument capable of unprecedented magnifications—had concentrated on measuring the zigzag movements of individual suspended particles. The observations were difficult to make and difficult to interpret. They suggested a paradox: the shorter the time interval under consideration, the higher the particle's speed seemed to be, tending toward infinitely fast. This was physically impossible, a perplexing enigma for the researchers.

Perrin took a different approach from the outset, studying the motion of the suspended particles in a statistical manner. He was able to verify that the vertical distribution of the suspended particles enters a statistical equilibrium with an exponential decrease in the particles' volume density. With these data, Perrin succeeded in determining Avogadro's number with an unprecedented precision: 6.8×10^{23} atoms per gram atom.

It was at this point that Paul Langevin, a professor of physics at the Collège de France, told Perrin about recent theoretical work done by Albert Einstein and Marian von Smoluchowski. In 1905, Einstein (Figure 8.7), who had been a technical expert third class at the Swiss Patent Office in Bern since 1902, submitted his PhD dissertation, *On a New Determination of Molecular Dimensions*, to the University of Zurich. He also sent two papers on Brownian motion to the *Annalen der Physik*, then a leading journal in the field.[5] Aiming to prove the existence of atoms and molecules with these studies, Einstein worked out a new approach that united statistical mechanics and thermodynamics. If statistical mechanics is valid, he reasoned, then any particle immersed in a "bath" of atoms or molecules must behave like a very large atom or molecule that is in thermodynamic equilibrium with its environment. Without realizing it, Einstein extended work begun by Louis Gouy, who in the late 1880s had posited that the irregular motions of the suspended particles stem from their collisions with a large number of invisibly small molecules of the surrounding fluid. In order to test the molecular hypothesis, Einstein proposed that experimenters should study the diffusion of suspended particles, that is, their displacement from their initial positions over time. The suspended particles, like Perrin's gamboge granules, would move rather slowly compared with the speed of the fluid's invisible atoms and molecules and would provide, through their motion, a magnifying glass into the world of atoms. If the molecular hypothesis is correct, Einstein argued, the average displacement of a suspended particle from its point of departure must grow with the square root of the time elapsed, the particle moving away twice as far on average from its initial position after a fourfold time interval has passed. This relation has since been called Einstein's square root law, and it provides another approach to determining Avogadro's constant. In passing, Einstein remarked that scientists had been right in studying Brownian motion to obtain new insights into the molecular hypothesis, but that in focusing on the particles' speed along the zigzag paths they had simply chosen the wrong quantity to measure. Marian von Smoluchowski, a professor of physics in Lemberg (Lvov) in present-day Ukraine and a leading expert on statistical physics, came to the same conclusions.

Figure 8.7 Albert Einstein during his tenure at the Swiss Patent Office for
Intellectual Property in Bern (1902). Courtesy of Archiv für Kunst und
Geschichte, Berlin.

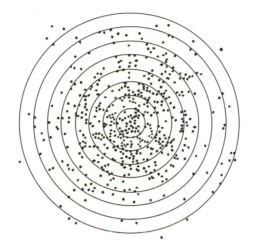

Figure 8.8 Einstein's theory of Brownian motion predicts that small particles in a suspension should, on average, be displaced from their initial position in proportion to the square root of the time that has passed. By measuring the displacement of individual gamboge particles after a certain time, Jean Perrin was able to test this theory, which was based on the assumption that matter consists of atoms and molecules. This plot shows the distribution of the relative displacements of 500 particles from their initial positions (at the circle's center) after 30 seconds, as measured by Perrin's student Chaudesaigues. The radius of the inner circle is 1.7 micrometers, and the diameter of each gamboge particle amounts to 0.2 micrometers. Reproduced from Jean Perrin, *Les Atomes* (Paris: Librairie Félix Alcan 1913), 169.

When Perrin learned of Einstein's theory, he realized immediately that he had made just the right kind of measurements to test the square root law. Einstein's prediction was neatly borne out (Figure 8.8), and when Perrin published his results in 1909, the world of physics was awestruck by this convincing demonstration of the reality of molecules. Perrin, however, went one step further toward making molecular reality an established scientific fact. By the year 1900, attempts to determine Avogadro's number had developed into an industry, with scientists deriving it from an amazing variety of physical and chemical phenomena. Yet if molecular reality was a universal phenomenon, Perrin reasoned, then all of these different approaches should yield the same number. Conversely, if independent determinations of Avogadro's constant converged toward one and the same number, this would be a convincing argument for the hypothesis of molecular reality.

From 1909 onwards, Perrin kept a running record of values obtained for Avogadro's number by himself and others. In hindsight, his publications demonstrate the amazing swiftness with which researchers zeroed in on a value of about 6×10^{23} atoms per gram atom. Measurements of atmospheric extinction were crucial in Perrin's endeavour, and were among the best values obtained at the time (Table 8.1). Deriving Avogadro's number from skylight measurements remained an active field of investigation throughout the early twentieth century. A suitable site for such work seemed to be Mount Wilson, just on the outskirts of Los Angeles, California. A large astronomical observatory had been built there with support from the Carnegie Foundation. It was from the top of this mountain that, in 1914, Charles Abbot and Frederick Fowle determined the number density of air molecules at sea level to be 2.69×10^{19} molecules per cubic centimeter. This value corresponds to an Avogadro constant of 6.02×10^{23} atoms per gram atom.

The more precise the measurements became, the more they converged to this number. Perrin's critics were silenced. In 1926, he was awarded the Nobel prize in physics. On this occasion his friend, Jacques Loeb, wrote to him:

> I hardly need to tell you that your proof of the real existence of molecules will live in science as an epoch-making discovery—and not only in science but in our general philosophical view of the universe.[6]

ON THE EXISTENCE OR NONEXISTENCE OF MOLECULAR SCATTERING

So far we have tacitly assumed, just as statistical mechanics does, that the atmospheric particles that scatter light are completely independent

TABLE 8.1

AVOGADRO'S NUMBER AS DETERMINED FROM A VARIETY OF DIFFERENT
OBSERVATIONS IN THE EARLY TWENTIETH CENTURY

Phenomenon	Avogadro's Number as Derived from the Respective Phenomenon (atoms per gram atom)
Viscosity of gases	6.2×10^{23}
	6.83×10^{23}
Brownian motion	6.88×10^{23}
	6.5×10^{23}
	6.9×10^{23}
Radiation of black bodies	6.4×10^{23}
Radioactivity	6.4×10^{23}
	7.1×10^{23}
	6.0×10^{23}
Critical point opalescence	7.5×10^{23}
Atmospheric extinction due to Rayleigh scattering	3.65×10^{23} (Lorenz 1890)
	5.93×10^{23} (King 1913)
	6.02×10^{23} (Fowle 1914)
Best value (1996)	6.022×10^{23}

SOURCES: Jean Perrin, *Les Atomes* (Paris: Félix Alcan, 1913), 289; Ludwig Valentin Lorenz, "Lysbevaegelsen i og uden for en af plane Lysbolger belyst Kugle," *Det Kongelige Danske Videnskabernes Selskabs Skrifter*, ser. 6, *Naturvidenskabelig og mathematisk Afdeling* 6 (1890): 1–62; Louis Vessot King, "On the Scattering of Light and Gaseous Media, with Applications to the Intensity of Sky Radiation," *Philosophical Transactions of the Royal Society of London* A212 (1913): 375–400; F. E. Fowle, "Avogadro's Constant and Atmospheric Transparency," *Astrophysical Journal* 40 (1914): 435–442; CODATA Bulletin 68 (1996): 963.

of each other. This condition seems necessary to assure the incoherence of scattering, which means that the phases of the individually scattered waves are randomly distributed. Only under this condition is the light scattered by N molecules N times more intense than the light scattered by a single molecule. All the properties of scattering discussed so far rely on this assumption. But if we give credence to the number densities of air molecules as determined around 1900, a surprising paradox arises. Not only are these molecules much smaller than the wavelengths of visible light, they also get very close to each other. At sea level the average distance between the air molecules is about one hundredth of the wavelengths of visible light. An incoming light ray should therefore be capable of exciting several neighboring molecules. On average, for each of these excited molecules there will be another excited molecule whose scattered light differs in phase by half a wavelength. As a result of destructive interference, waves with the same wavelength but a phase shift of half a wavelength mutually extinguish each other (see Figure 6.6). This implies that the intensity of the resultant scattered light should be nil, even though a large number of particles are scattering light. Consequently, the atmosphere should be completely transparent (and the daytime sky black), even when air molecules scatter sunlight, contrary to our everyday experience.

This bizarre paradox was resolved between 1908 and 1910 by Einstein and Smoluchowski. Both were interested in critical point opalescence, a phenomenon that at first seemed unrelated to the color of the sky. At specific combinations of temperature and pressure, many chemical substances can occur alternatively as a gas, a liquid, or a solid. This is called their critical point. Although they are largely transparent as liquids or gases, the optical turbidity in these substances may increase considerably as they approach the critical point. To account for this "opalescence," Smoluchowski and Einstein proposed a theory of light scattering at density fluctuations. They realized that this theory could be applied to the situation in the atmosphere, even though its gases are far from the critical point. For instance, molecular nitrogen reaches its critical point at $-146°C$ and thirty-three times the sea-level atmospheric pressure.

Let us imagine a large number of light-scattering molecules in the atmosphere. Statistical mechanics predicts that the distances between the molecules vary constantly, and that they often collide. Because of these microscopic motions, their number in a unit of volume is subject to statistical variation. We can assume the motion of individual molecules to be independent of the others and, given this condition, Poisson statistics aptly describes the fluctuations of the particle number in a volume. It says that, with a probability of two thirds, a volume containing N molecules on average will, at any given moment, contain $N \pm \sqrt{N}$ molecules, where $\sqrt{N}$ describes the fluctuation around the average number. If we believe a volume to contain exactly 100 molecules, there are actually about 100 ± 10 molecules in it. The number will most likely be somewhere between 90 and 110.

Let us now consider Figure 8.9. It shows two volumes of air that are independent of each other but aligned perpendicular to the incoming solar rays, differing in their distance from an observer by half the wavelength of the incoming light. If the number of molecules were exactly equal to N in both volumes, all the scattered light would be mutually extinguished, since every scattering molecule would have a counterpart half a wavelength away. However, due to the fluctuations, the number of scattering molecules in the two volumes will differ by about $\sqrt{N}$. Since the number N is very large, we shall pretend that $\sqrt{N}$ is an integer number. It does not matter which one of the two volumes contains more and which one fewer molecules. In either case there will be $\sqrt{N}$ molecules whose scattered light is not extinguished via destructive interference by molecules in the other volume. If we let A be the amplitude of the light scattered off a single molecule, the sum of the amplitude of the light scattered by $\sqrt{N}$ molecules will be the product $\sqrt{N}$ A. Since the total intensity of scattered light, $I,$ is the square of the amplitude, we have $I \approx NA^2$. This is exactly the result we would have obtained by adding up the intensities for incoherent molecular scattering. Thus the fluctuation theory shows that all the features of Rayleigh scattering conform to incoherent scattering, despite the individual molecules being so close to each other. However, in this view we may consider the fluctuations as the scatterers, rather than the individual particles themselves.

While fluctuations in the number density of gaseous air thus "save" blue skylight, liquids and solids have some interesting light-scattering

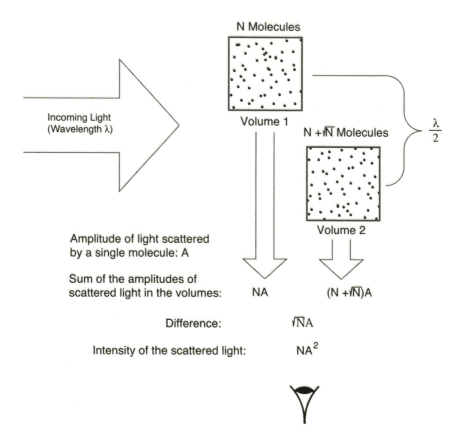

Figure 8.9 Two volumes of air are illuminated and scatter the incoming light. If the number of scatterers in volumes 1 and 2 were the same, all laterally scattered light would be canceled out due to destructive interference, since these volumes are half a wavelength apart. All of the net scattering results from the difference in the numbers of scatterers in the two volumes, that is, from the fluctuations in the number densities.

properties of their own. In these media the particle number per unit volume is subject to fluctuations that are smaller than those found in gases, and Poisson statistics does not apply. Although there are very many potential scattering molecules in liquid water, it is much more transparent than a volume of a gas with the same number of molecules. Many liquids and solids, such as water, ice, and glass, are transparent because in them scattered light is diminished via destructive interference. Light scattering occurs in these materials even far from the critical point, but its net intensity is generally low.

In his 1910 paper on critical point opalescence, Albert Einstein, now a professor of theoretical physics at Zurich University, does not treat the molecules themselves as subject to statistical variations within the medium, but rather the index of refraction. He considers the matter itself as continuous. On these grounds he succeeds in deriving the "inverse fourth power" law, as previously known from Rayleigh's theory, and concludes:

> As shown by a rough calculation, this formula is capable of explaining the existence of the predominantly blue light emitted by the illuminated sea of air. Here it is remarkable that our theory does not make *direct* use of the assumption of a discrete distribution of matter.[7]

Does this mean that there could be a blue sky without the air molecules? The answer is no. Einstein rightly emphasizes the word "direct." After all, it is the molecules that produce the fluctuations in the index of refraction, and as such they are necessary for scattering the light. It is remarkable that Einstein's derivation yields the same law, even though his approach was completely independent of Rayleigh's. This is a nice demonstration of the consistency of the physical worldview. While the work of Einstein and Smoluchowski on light scattering via fluctuations reaffirms the amount of Rayleigh-scattered light mathematically, it also reminds us that deriving formulas is one thing, but interpreting them is quite another.

THE BLUE SEA

Much as John William Strutt had done forty years earlier, Einstein and Smoluchowski contemplated the scattering of light *waves*, thus treating light in terms of classical, nineteenth-century physics. Einstein did so despite having been instrumental in developing the rather different concept that light consists of photons, that is, quanta of specific energies. He had introduced this concept in 1905 in order to explain the photoelectric effect (work for which he won the 1921 Nobel prize in physics), but it soon proved successful in explaining a large variety of physical phenomena, including the heat capacity of solids, and the radiation emitted by them. His reasoning has become a cornerstone of modern quantum physics. In our context, the question is whether light scattering, when treated in terms of photons rather than light waves interacting with the scattering particles, accounts for the observable properties of skylight. Answering this question requires a detour to another blue riddle, that of the color of the sea.

In 1872, the honeymooners John William Strutt and his wife Evelyn crossed the Mediterranean on a steamship bound for Egypt. We have already come to know Strutt, the soon-to-be third Lord Rayleigh, as a scholar who was easily distracted from romance by his various studies. The year before, he had studied the sky's blueness. Aboard the ship, it was the color of the sea that caught his attention. Was it due to the same process of light scattering as the color of the sky? Close scrutiny convinced him that the color of the sea was mostly a reflection of the blue sky, with an additional contribution from light absorption by matter floating in the water.[8]

Half a century later, this interpretation was challenged when another great scholar crossed the Mediterranean on a steamship. In September 1921, Chandrasekhara Venkata Raman, then a professor of physics at Calcutta University in India, returned from a visit to England. Raman (Figure 8.10) had become known for investigating the acoustics of Indian musical instruments such as the string instruments *tambura* and *veena,* as well as the *tabla* drum, studies he did in his spare time while employed as a financial officer in the British colonial administration. In mid-1921 he had represented his university at an international University Congress of the British Empire in Oxford and had made use of the opportunity to visit colleagues in Cambridge and London. Raman's voyage home aboard the steamship Narkunda departed from Southampton, then followed a course around Spain, through the Strait of Gibraltar, across the Mediterranean, into the Red Sea via the Suez Canal, past

Figure 8.10 Chandrasekhara Venkata Raman in a photograph taken in 1948. Courtesy of Bildarchiv Preussischer Kulturbesitz, Berlin.

Aden into the Indian Ocean, and reached Bombay two weeks after leaving England. Like Strutt, Raman was fascinated by the deep blue color of the Mediterranean and began to ponder its origin. Raman carried with him a pocket spectrograph, which he used to observe the surface of the water. Soon he was convinced that Rayleigh's explanation of the sea's color was wrong, and that the water molecules were capable of scattering light in much the same way as the air molecules, according to the same inverse-fourth-power law that Rayleigh had derived in 1870. Raman's voyage across the Mediterranean took place ten years after Einstein and Smoluchowski had investigated scattering at density fluctuations. Knowing of their findings, he was not too surprised to find that the sea exhibited this phenomenon.[9]

Raman was intrigued by the notion that fluids are capable of scattering light. Upon returning to Calcutta, he undertook a series of detailed laboratory studies of this phenomenon that would occupy him off and on for the next seven years. Eventually, in late February 1928, Raman and his assistant K. S. Krishnan observed something remarkable. Although most of the scattered light was of the same wavelength as the light illuminating a fluid, they noticed that a small fraction was scattered with a wavelength longer than that of the incident light. It soon became clear that this was not an isolated occurrence; the same effect was observed in each of the more than sixty different fluids studied by Raman and Krishnan. This phenomenon was quite unlike Rayleigh scattering, in which the wavelength of light remains unchanged. Rayleigh scattering may be termed elastic; that is, all the incident energy bounces off the molecules' bound electrons almost instantly. In contrast, Raman scattering, as the new phenomenon was soon named, is inelastic, and part of the energy of the incident light is converted into rotational and vibrational modes of the molecules. This type of scattering can be understood only in the framework of quantum mechanics; it had been predicted in 1924 by the theoreticians Hendrik Kramers and Werner Heisenberg. Raman scattering builds a bridge toward explaining phosphorescence (see Chapter 7), in which the energy of incoming light is emitted at a longer wavelength, perhaps after being stored in a molecule for some time.

It turns out that the blue color of the sea is due to a combination of scattering, reflection, and absorption. As such, neither Rayleigh nor Raman arrived at the correct solution alone, but once their insights are joined together, the riddle is solved.

LIGHT SCATTERING ABROAD

The blueness of the sea extends light scattering into a realm different from Earth's atmosphere. And indeed, while the circumstances leading to the blueness of the daytime sky may seem unique, with particles smaller than the wavelength of light being illuminated in front of a dark background, Rayleigh scattering can be observed far beyond Earth's atmosphere. We can look for it in any place where its basic ingredients are found.

Perhaps the place where astronomers have studied light scattering most intensely is in the atmospheres of planets and stars. Considering that Earth is a planet, it should not surprise us that Rayleigh scattering occurs in other planetary atmospheres as well. It does occur in the atmospheres of the planets Venus, Mars, Jupiter, Saturn, Uranus, Neptune, and Pluto, even though not all of these appear blue. Careful examination shows that planetary atmospheres differ greatly in their composition and density, so that absorption, as well as single and multiple scattering, conspire to create a variety of different colors. (Chapter 10 deals with this in depth.) Rayleigh scattering exerts an important influence on the appearance of the rings of Saturn as well; without it, the polarization seen in them would be inexplicable.

Stars are huge balls of gas that are fueled by nuclear reactions in their cores, the sun being our nearest example. It may seem surprising that stars have atmospheres. But this is actually a matter of definition: astronomers call the layers of a star from which we receive direct radiation a stellar atmosphere. Those layers have held great significance in astronomy, for all our knowledge of the stars depends on the study of the light that we receive from them—that is, on measurements of its direction and composition. It is in the atmospheres that the stars' chemical composition and the physical state of their surface are fingerprinted onto the light that is emitted. Astronomers have managed to describe the transfer, emission, and absorption of this radiation in a rather satisfactory way. It turns out that Rayleigh scattering is quite significant in the extended, cool atmospheres of giant stars, but less so in stars like the sun.

Another major site for light scattering is the space between the stars. Not at all devoid of matter, this space is filled with a mixture of gas, hot plasma, and cold dust. Light scattering in this interstellar medium is common and leads to a pronounced extinction of the light from remote stars in our Milky Way galaxy. Even the light from beyond our galaxy is scattered and absorbed there in a way that often troubles astronomers, but it also provides them with much insight into how our cosmic surroundings are composed and how galactic matter is recycled. Rayleigh scattering is but one of several processes involved in interstellar extinction. It turns out that most of the scattering and absorption in the Milky Way is due to particles that are as big as, or even bigger than, visible light waves. Thus, Mie scattering, absorption, and reflection also play prominent roles. Photographs of gas nebulae in our galaxy show that many of them shine with a sky-blue light, the most famous example being the Pleiades star cluster in the constellation Taurus, a common sight in the winter sky. However, most of these nebulae, including the one surrounding the Pleiades, are reflection nebulae. These nebulae *reflect* the light of nearby stars; only a small fraction of their light is due to Rayleigh scattering.

Though hundreds of light years away, to an astronomer these nebulae are nearby objects. It turns out that Rayleigh scattering affects even the most distant phenomenon known to us. This is the cosmic microwave background, a feeble radio signal that reaches us from all directions in the sky. There is a consensus among astronomers that this radiation originated in the era of recombination, 400,000 years after the universe formed in the Big Bang, 13.7 billion years ago. Initially filled with hot, opaque plasma, the universe had been cooling since its formation, the but at this point in its history it became transparent to radiation. It had become cool enough for electrons and atomic nuclei to combine to form the elements hydrogen and helium. The transition resembles the surface of a cloud in Earth's atmosphere. Inside a cloud, light is scattered by the many small droplets and keeps changing its direction of propagation, yet after leaving the cloud, light may propagate unscattered for much longer

Figure 8.11 John William Strutt, the third Lord Rayleigh, in a portrait from 1905. Reproduced from Robert John Strutt, *Life of John William Strutt, Third Baron Rayleigh* (London: Edward Arnold & Co., 1924), frontispiece.

distances. With this picture in mind, what we see in the cosmic microwave background is aptly called the surface of last scattering. As the
word microwave in its name implies, this radiation is not visible to the
naked eye, but is best observed with radio telescopes. While the surface of
last scattering is dominated by the scattering of free electrons by atoms
(Thomson scattering), Rayleigh scattering must have occurred as well.
This was argued in 1970 by James Peebles and J. T. Yu, both then at
Princeton University. Thirty years later, in 2001, their colleagues
Qingjuan Yu, David Spergel, and Jeremiah Ostriker realized that current
satellite missions, in particular the NASA satellite WMAP (in operation
since 2001) and the European Space Agency's Planck mission (launch
scheduled for 2007), may be able to observe the signature of Rayleigh
scattering at cosmic recombination in the microwave background.[10]

Quite literally, Rayleigh scattering has come a long way since John
William Strutt first discovered it. The physicist could not have foreseen
that the process of light scattering he conceived of in that fall of 1870
would turn out to be such a far-flung phenomenon, not even as he lay
on his deathbed in 1919, almost half a century after his remarkable
achievement. Yet even if he had known it, he would probably have remained the modest man who, after being awarded the Order of Merit in
1902, said of himself:

> The only merit of which I personally am conscious is that of hav-
> ing pleased myself by my studies, and any results that may have
> been due to my researches are owing to the fact that it has been a
> pleasure to me to become a physicist.[11]

CHAPTER 9

Ozone's Blue Hour

I t can literally take your breath away to stand on a mountaintop and gaze up at the twilight sky. As a graduate student in astronomy, I discovered this while contemplating distant galaxies from an observatory on Capilla Peak in New Mexico (Figure 9.1). The adventure began after an hour's drive from Albuquerque, when we turned onto the dirt road leading to the mountain. Randy Grashuis, the expert telescope operator, drove the four-wheeler, swerving around pot-holes and over cattle guards, and winding up through the forest of pine, spruce, and aspen trees lining the steep slopes to the 9,300-foot summit. The late afternoon view from the top was simply magnificent. As soon as Randy turned off the motor, a quiet serenity spread out around us. The clean, dry air smelled of pine trees. The mountains nearby rose up from the surrounding plains like islands out of the ocean. In the west, the setting sun was approaching the horizon. Far beyond was the Rio Grande, just barely visible as a reflection of light from the wide valley below. To the north were the Sangre de Cristo Mountains near Santa Fe. Eastward the plains extended far into the central United States. Nearby, to the south, Capilla's twin peaks were visible. Free of all obstructing buildings and power lines, the cloudless sky seemed vaster and its air clearer than I had ever seen before.

It was almost twilight, and the telescope in the observatory's dome still had to be prepared for the night. We opened the dome slit, started the computer system, and inspected the telescope and its electronic camera. Once these tasks were done, I had time for a stroll outside and another look at the sky. The sun had now set, leaving behind a yellowish afterglow. Above and all around, the sky was radiating a deep, brilliant blue that was quite different from the hue of the daytime sky (see Color Plate 21). This blue reminded me of stained-glass windows in a medieval

Figure 9.1 Capilla Peak Observatory, situated on a 9,300-foot peak overlooking the Rio Grande, is run by the University of New Mexico. The dome houses a twenty-four-inch telescope. Courtesy of William A. Miller.

cathedral. We seemed to be standing inside a huge sapphire. Slowly the stars began to appear, one by one (see Color Plate 22).

I have seen this blue many times since then. It isn't found just on remote mountaintops. Rather, it is an everyday phenomenon, visible to anyone who is paying attention. Many poets have. The "blue hour" is what they called the moment when "the sky has lost the sun, but has not yet gained the stars," as Jacques Guerlain wrote in 1911.[1] Poets and novelists have associated it with emotions of unfulfilled desire, melancholy, and transcendence. William Least Heat Moon, who in his book *Blue Highways* tells of his epic journey along the back roads of North America, describes the blue hour as the moment when he most felt the pull of the road, urging him on.[2] Heat Moon passed by not far from Capilla Peak.

Another writer, Gottfried Benn, wrote poems about the blue hour, the feelings of happiness, danger, and illusion associated with it: was it real or just a dream? The perfume maker Guerlain tried to create a perfume that evoked these moods.

While scientists may share these emotions, for them the blue hour poses a riddle. Light scattering is sufficient to explain the blue of the daytime sky and the red of the setting sun, but it fails to explain why the sky remains blue after sundown. By that time, all the rays of sunlight that contribute to the skylight should have traversed the atmosphere on their long paths, losing their blue, short-wavelength components on the way due to scattering. Only those rays that skim along the upper reaches of the atmosphere would scatter blue light toward an observer. However, the number of scattering molecules present at high altitudes is too small and the light too weak for the sum of all the scattered light to add up to a visible blue. If you only take scattering into account, the twilight sky should look yellowish or greenish. But we all see a blue sky over our heads.

Forty years before Randy and I went to Capilla Peak, and 150 miles farther south, the secret behind this deep blue light was revealed. It is a visible trace of the life-protecting ozone layer in the stratosphere, about 25 to 35 kilometers (15 to 20 miles) above Earth's surface. This surprising discovery was made by scientists from the United States Naval Research Laboratory. Led by Edward Hulburt, the laboratory's research director, they had gone to Sacramento Peak in southern New Mexico to observe the twilight sky and study Earth's upper atmosphere. Like Capilla Peak, Sacramento Peak is far away from city lights, industrial pollution, and haze, all of which could disturb the sensitive measurements Hulburt needed to make. After struggling in vain to interpret their data, he realized that ozone was needed to account for the brightness and color of the twilight sky. This is because the ozone layer acts as a huge color filter, blocking out the orange and red components of scattered sunlight and leaving only the blue behind. Here was another hint that our knowledge about the lower atmosphere might not be valid higher up. After all, ozone is barely present near the ground.

A STRANGE-SMELLING GAS

For science, the story of ozone begins in 1839. This was the year when the Swiss chemist Christian Friedrich Schönbein noticed that a third gas formed during the electrolysis of water, in addition to hydrogen and oxygen. Since the gas made its presence known by a uniquely pungent smell, Schönbein named it ozone (from the Greek *ozein*, to smell like something). The smell resembles that of chlorine and is often encountered near photocopy machines, sun-ray lamps, or laser printers.

Twenty-five years passed before Jacques Louis Soret discovered that this gas was composed of molecules consisting of three oxygen atoms bound together (Figure 9.2). Its chemical formula is O_3. A French chemist, Jean Auguste Houzeau, had discovered in 1858 that atmospheric air contains a trace of ozone.[3] This came as a surprise, because Schönbein had deduced from its chemical behavior that the gas was a strong oxidant that quickly reacted with other substances, and thus was rapidly destroyed. It is this oxidizing effect of ozone that is dangerous for humans when they come in direct contact with the gas.

In 1878 Alfred Cornu made an intriguing discovery that at first glance did not seem to have anything to do with ozone. Cornu had been investigating the ultraviolet spectrum of the sun, which is invisible to us. He observed that the ultraviolet light abruptly broke off at about 300 nanometers; no rays could be detected below this wavelength. Cornu speculated that this might be due to a peculiarity of the sun. Perhaps it does not emit any rays with wavelengths this short. Another possibility was that Earth's atmosphere swallowed up these energy-rich ultraviolet rays. After observing the solar spectrum at different times of day, Cornu found an answer to this question (Figure 9.3). The distance that sunlight travels through the atmosphere depends on the height of the sun above the horizon. He noticed that the wavelength at which the ultraviolet spectrum discontinues increases as the length of the path of sunlight increases. Thus, the discontinuity in the solar spectrum did not point to a peculiarity of the sun but was an indication that ultraviolet light was absorbed on its way through the atmosphere.[4]

Oxygen Molecule

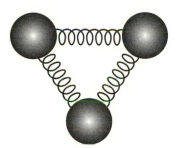

Ozone Molecule

Figure 9.2 Structure of molecules of atmospheric oxygen (O_2) and ozone (O_3). The forces between the atoms are depicted as springs.

Two years later the Irish chemist Walter Hartley proposed a likely agent for the absorption: ozone. In his laboratory, Hartley had studied the absorption of light by different gases. He noticed that extremely small amounts of ozone could substantially reduce the incident radiation with wavelengths below 300 nanometers. This was exactly the lower boundary that Cornu had found for the atmospheric absorption. Hartley speculated that the absorption could be caused by a layer of ozone located high up in the atmosphere. But he had no direct proof. The deep absorption features in the ultraviolet range of the spectrum later came to be called Hartley bands.

Without realizing it at the time, William Huggins found another absorption band in the ultraviolet spectrum of ozone. In 1890, Huggins, an

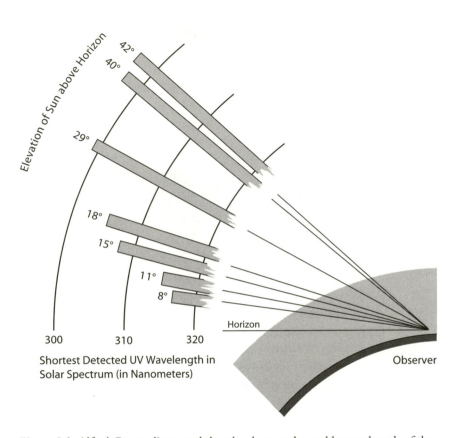

Figure 9.3 Alfred Cornu discovered that the shortest detectable wavelength of the sun's ultraviolet spectrum depends on its height above the horizon. At noon he could detect the spectrum down to 295 nanometers. Shortly before sundown, however, only longer wavelengths of ultraviolet light (from 317 nanometers upward) reach Earth. This was the first proof of the atmosphere's ability to absorb the sun's short-wave ultraviolet radiation.

amateur astronomer who ran a mercury plant for a living, made the first photographs of stellar spectra. To Huggins, the absorption lines visible in the photographs were fingerprints of the chemical elements that made up the stars. These lines agreed with the spectra of known elements he had investigated in his laboratory earlier—a clear indication of the stars' chemical composition. Huggins realized that the stars were essentially lustrous spheres composed of different gases. However, there were several lines in the spectra that he could not explain. These included a number of dark absorption lines between 319 and 334 nanometers in the spectrum of Sirius, the brightest star in the night sky. Almost thirty years later, Alfred Fowler and Robert John Strutt (the son of John William, the third Lord Rayleigh) compared the lines observed by Huggins with the spectrum of ozone. They realized that what Huggins had seen were ozone absorption lines from Earth's atmosphere. Today they are called Huggins bands. Like the Hartley bands, they are invisible to the human eye.

The French chemist James Chappuis seems to have been the first ozone researcher to notice its blue color. He attributed this blue to the gas's absorption of yellow, orange, and red light rays. This feature is now called the Chappuis absorption bands. Since the presence of ozone in the atmosphere had already been proved by Houzeau, Chappuis suspected that it might be involved in creating the blue color of the sky.[5] However, he knew that absorption does not generally cause light to become polarized. Therefore, since the light of the clear sky is blue and polarized, some other process must be involved as well. Rayleigh scattering, for example.

Until 1918, no one knew how much ozone was present in the atmosphere. But in that year, thanks to new advances in spectroscopy, Charles Fabry and Henri Buisson were able to calculate the ozone concentration. Their measurements confirmed that Houzeau had been right in claiming that only traces of ozone were present in the atmosphere. If the entire atmosphere were subjected to the pressure at sea level, it would be just 10 meters thick. Ozone is such a minor component of the atmosphere that, under those conditions, it would contribute a layer only three millimeters thick. According to Fabry and Buisson, this small amount must occur somewhere in the lower 50 kilometers of the atmosphere.[6]

Studies of ozone were fashionable in those days. Only a few years later, building on the work of Fabry and Buisson, Gordon Dobson developed an easy-to-use ozone meter. Routine measurements of ozone were now possible, and Dobson, a meteorologist from Oxford University, was

instrumental in establishing a worldwide network of stations for this purpose. Many of these measuring stations are still in operation today. In
recognition of Dobson's contribution, the unit for the total amount of
ozone in a column of air is named after him: one Dobson unit corresponds to a thickness of 0.01 millimeters of ozone under so-called normal conditions—that is, sea-level air pressure at a temperature of 0°C.
Thus, the amount of ozone measured by Fabry and Buisson equaled 300
Dobson units.

Clues to the precise location of the ozone layer posited by Hartley,
and later by Fabry and Buisson, were found in 1929 by Paul Götz. As the
head of one of Dobson's ozone stations in Arosa (Switzerland), Götz developed a clever method that uses simple observations of Chappuis absorption to determine the distribution of ozone at different altitudes.
During an expedition to the Arctic archipelago Spitsbergen, Götz was
able to make the necessary measurements. He found that the greatest
concentration of ozone in the stratosphere was located 25 kilometers
(15 miles) above Earth's surface. Balloon ascents and missile flights that
have explored the upper atmosphere since the 1930s have confirmed
this finding. These observations led to another surprising discovery:
while the air temperature at first drops steadily with increasing altitude,
it rises significantly near the ozone layer, and then continues to fall again
above it (Figure 9.4).

A PROTECTIVE SHIELD

Around 1930, there were numerous indications that ozone is a unique
component of Earth's atmosphere. It absorbs the energetic ultraviolet
rays of the sun, may warm the stratosphere, and possibly contributes to
the blue color of the sky. But most intriguing of all is the fact that this
gas is present even in small amounts in the atmosphere, since it reacts so
quickly and is destroyed in the process.

The British geophysicist Sydney Chapman suggested a solution to
this riddle.[7] He posited that the ozone molecules in the stratosphere do
in fact frequently react with other molecules and are consumed in the

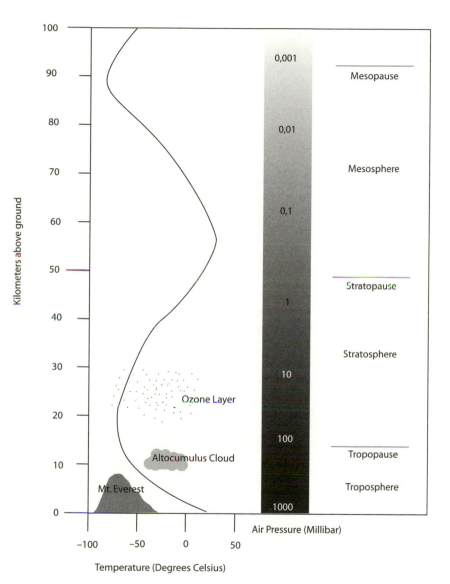

Figure 9.4 Structure of the lower atmosphere.

process. However, at the same time an equally large number of ozone molecules are being produced by the ultraviolet light of the sun. These two processes result in what is now called a photochemical equilibrium. According to Chapman, the ozone layer is in equilibrium only if a large number of oxygen molecules are present in the upper atmosphere. In the daytime, the oxygen molecules are irradiated by sunlight, which contains ultraviolet, visible, and infrared radiation. Energetic solar ultraviolet rays (those with wavelengths below 230 nanometers) are sufficient to dissociate the oxygen molecules (O_2). In other words, oxygen molecules are split into two oxygen atoms:

$$O_2 + UV \text{ light} \rightarrow O + O.$$

Each of the oxygen atoms thus released can bind to an oxygen molecule and form ozone. This ozone, however, is highly unstable and dissociates very quickly, unless a third collision partner is available to divert the energy that is liberated. This partner, which we shall call M for now, can be an atom or a molecule:

$$O + O_2 + M \rightarrow O_3 + M.$$

Since two atoms are released when an oxygen molecule is split, the same reaction must take place twice. The sum of these two synthesis reactions gives us

$$3 \ O_2 + UV \text{ light} \rightarrow 2 \ O_3.$$

Thus, if illuminated by ultraviolet radiation three oxygen molecules can produce two ozone molecules. However, the oxygen molecule with two atoms is not the only one that is sensitive to (ultraviolet) light. Even lower-energy, visible sunlight suffices to dissociate the ozone molecule. In the process, an oxygen molecule (O_2) and an oxygen atom (O) are set free. Since the chemical bonds that hold the ozone molecule together are weaker than those of the oxygen molecules, visible light and near-infrared light are sufficient to split it:

$$O_3 + light \rightarrow O + O_2.$$

The oxygen atom that has been released then seeks another binding partner. If this partner is an oxygen molecule, ozone will be formed again. Thus, a so-called circular reaction takes place. The binding partner could also be an ozone molecule. In this rarer case two oxygen molecules result:

$$O + O_3 \rightarrow 2\ O_2.$$

The sum of these two decomposition reactions is three oxygen molecules:

$$2\ O_3 + light \rightarrow 3\ O_2.$$

The fission of two ozone molecules releases three oxygen molecules. Thus, in a way the decomposition reaction is the reverse of the synthesis reaction: the equilibrium of stratospheric ozone is determined by the rates of formation and destruction of this gas. Every atmosphere containing oxygen, even on other planets, will be characterized by such a cycle, provided it is sufficiently exposed to intense ultraviolet radiation. In the case of Earth's atmosphere, the amount of ozone has apparently settled at a quite small value; there are only about ten ozone molecules for every million air molecules in the stratosphere.

Chapman's theory was an important milestone in ozone research, but it does not answer every question. It predicted that the amount of ozone should be five times larger than it is. To precisely account for the amount of ozone actually observed, we also have to consider the other substances and chemical reactions found in the atmosphere. The quantitative prediction of ozone concentrations is a difficult task, involving hundreds of equations, lengthy codes, and a lot of computation time on large mainframe computers.

During the fission of oxygen and ozone molecules, the energy of ultraviolet radiation is absorbed and transformed into heat. In this way, the ozone layer acts as a filter which blocks the high-energy ultraviolet

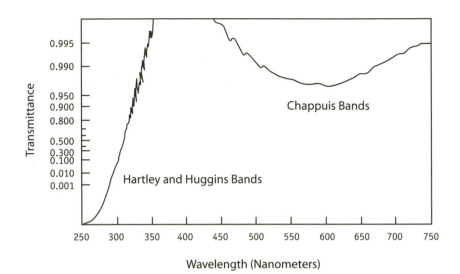

Figure 9.5 Absorption spectrum of ozone in the ultraviolet and visible ranges. The diagram shows the transmittance of a one-centimeter layer of ozone at sea level pressure and 0°C (0 = fully absorbing, 1 = fully transmitting). Reproduced from S. Bakan and H. Hinzpeter, "Atmospheric Radiation," in *Landolt-Börnstein, Numerical Data and Functional Relationships in Science and Technology*, Group V, vol. 4b, *Meteorology: Physical and Chemical Properties of the Air* (Berlin: Springer-Verlag, 1988), 140.

radiation of the sun. The Hartley and Huggins bands show the inherent ability of ozone to absorb ultraviolet light in its spectrum (Figure 9.5). This is why ozone is important for life on Earth. Since all living creatures are very sensitive to ultraviolet light, they must be protected from its effects if they are to live on the planet's surface. The energy of ultraviolet light is sufficient to damage cells and cause changes in the genetic material, the DNA. Skin cancer and cataracts are among the known dangers of this radiation. The history of Earth suggests that the formation of the ozone layer was most likely the prerequisite for evolved living creatures to leave the ocean. Since water absorbs ultraviolet light, bodies of water had long been the only safe haven from radiation. However, once the stratospheric ozone concentration reached its present level, about 500 million years ago, the land could be safely colonized.

A COLOR FILTER IN THE SKY

When James Chappuis suggested in 1881 that ozone might be involved in the color of the sky, he was not taken seriously. At that time physicists were enthusiastic about Lord Rayleigh's theory of light scattering. It seemed to easily account for the color and polarization of skylight. Thus, the idea of an ozone-blue sky quickly sank into oblivion. In the early 1950s, Edward Hulburt was certainly unaware of it when he began contemplating the twilight sky, and he did not suspect anything peculiar behind its colors. Rayleigh's theory seemed to have long since resolved the ozone issue. But then Hulburt (Figure 9.6) stumbled across it while comparing two sets of data gathered by research groups from the Naval Research Laboratory.

Measurements of the light from the twilight sky on Sacramento Peak made up the first set. They covered the different phases of twilight, beginning before sunset and extending well into the dark night. Taking such measurements was a daunting task, since the brightness of the zenith drops by a factor of seventy million between noon and the beginning of astronomical night. A suitable instrument needed to cover such a wide range and still be sensitive enough to measure the barely perceptible, feeble light of late twilight. A sensitive electric photomultiplier tube was used, yielding the most precise measurements of the kind to date.

Hulburt had been interested in such observations since the 1930s, because they seemed to be the best way to study the density and temperature of the upper atmosphere. The technique he used relied on the fact that as twilight advances, the sun sinks further below the horizon and illuminates only the upper part of the atmosphere. The lower layers have already sunk into the curved edge of Earth's shadow. If we assume that all the observed light from above is due to Rayleigh scattering, the number of scatterers at different altitudes can be calculated from the measured brightness of the twilight sky at different times. By 1938 Hulburt had derived the formulas involved in such a calculation.[8]

Now he was able to check these initial findings against a special set of independent data: direct measurements of the upper atmosphere. This

Figure 9.6 Edward Olson Hulburt in a photograph taken in 1956. Courtesy of the Optical Society of America.

was an exciting new field of research that had opened up immediately after World War II. A number of V-2 rockets had been captured in Germany and brought to the United States to allow American scientists gaining experience in rocket technology to also study the upper atmosphere. The rockets routinely reached altitudes of 160 kilometers (100 miles). As the Naval Research Laboratory's research director, Hulburt led the transformation of warheads into "specialized short-lived laboratories," as he wrote in 1947.[9] With these rocket laboratories, it was possible to measure the pressure and density of the air, as well as the incident solar radiation at altitudes far above the ozone layer. The rocket soundings were conducted at White Sands, a military testing ground a mere 80 kilometers (50 miles) from Sacramento Peak.

The next step was to compare the twilight data with the rocket measurements. Hulburt calculated the expected brightness of the sky at Sacramento Peak from the rocket measurements. To his surprise, the predicted twilight sky was two to four times brighter than the photomultiplier data had shown. This difference far exceeded the experimental margin of error. And what was worse, when he worked out the expected colors of the twilight sky, he realized that the observed sky was not only too bright but also the "wrong" color. Rayleigh's theory alone predicted that the overhead sky should change to a blue–greenish gray at sunset and turn yellow during twilight. This flatly contradicted all human experience that the overhead sky remains blue during twilight, and that its hue changes only slightly.

What had gone wrong? It was unlikely that Hulburt's calculations were in error. After all, they involved little more than Rayleigh scattering, which by then had stood the test of time. But perhaps the calculations were incomplete. Was there some other absorbing constituent of the air, which he had overlooked? Hulburt quickly realized that there was: ozone. The concentrations of this gas had been measured by the sounding rockets, and scientists from his laboratory had analyzed the data. Indeed, during the twilight measurements on Sacramento Peak, a nearby rocket launch had revealed that the total ozone over New Mexico amounted to about 240 Dobson units or, at sea-level pressure, a thickness of 2.4 millimeters. Hulburt knew that ozone was visible due to the absorption bands discovered by James Chappuis. Thanks to a recent study by Arlette and Étienne Vassy, the absorption features were now known with great precision. This French husband-and-wife team of chemists noted that the maximum absorption occurred at about 600 nanometers, just as Chappuis had anticipated. Thus, in the visible spectrum ozone primarily absorbs orange light. It also absorbs yellow and red light, but much less efficiently. The shorter wavelengths of blue light are hardly affected. This is how the Chappuis bands make the gas look blue.

The efficiency of ozone in absorbing visible light turned out to be more than a thousand times less than its efficiency in absorbing ultraviolet radiation. Thus, because of the Hartley and Huggins bands, a layer of ozone that would be only 0.07 mm thick at sea level absorbs 90 percent of the incident ultraviolet radiation below 300 nanometers. In comparison, the Chappuis absorption is so weak that it would take a layer 16.7 cm thick to absorb the same fraction of orange light. These numbers explain why the stratospheric ozone cannot make a substantial contribution to the blue color of the daytime sky. After all, at that time it would be equivalent to a layer just two or three millimeters thick under conditions of sea-level atmospheric pressure. As such it would absorb less than 2 percent of the incident orange light.

Taking these features into account, Hulburt corrected his calculations of the brightness and color of the twilight sky. Now everything fit together. The optical data from Sacramento Peak agreed with the rocket data from White Sands. The theoretical twilight sky was as dark as the observed sky. And it remained blue. Hulburt was surprised by the strong effect of ozone absorption on the color of the sky. While imperceptible in the daytime, it quickly accounts for two-thirds of the observed "blueness" at sunset. At twilight, all the blue results from Chappuis absorption.

A look at the basic geometry involved in the sunset shows why this is so. Due to their oblique incidence, the light rays pass through a longer stretch of the ozone layer, and this intensifies the effect of the Chappuis absorption. Now the bands can filter out up to 40 percent of the orange light—enough to leave a clearly visible trace in the sunset sky which becomes dominant when the sun is below the horizon. This selective absorption reduces the brightness of the twilight sky and causes a color shift toward the shorter wavelengths. Thus, the ozone in the stratosphere functions like a color filter that covers the entire sky, making the sky over our heads blue long after the sun has set.

Rayleigh scattering and Chappuis absorption are entirely different processes, but they both separate the blue rays from incident sunlight. As we saw in Chapter 7, in Rayleigh scattering small particles are briefly excited and immediately re-radiate the energy they have received. Since this is more likely to happen at short wavelengths, the scattered light is predominantly blue. The scattering particles are left intact, regardless of whether they are molecules or small aerosols. In contrast, Chappuis absorption results when the ozone molecules break apart. When a molecule of ozone is hit by a photon of visible sunlight, it receives more energy than it can hold. The molecule is set into a stretching mode of vibration, in which one or two of its three oxygen atoms move back and forth. Typically, the energy is too large to be contained in the molecule and the molecule breaks apart after one or two oscillations. This is an extremely brief process, lasting only a few pico seconds (one pico second is a millionth of a millionth of a second). The short duration of this dissociation process, as it is called, is seen in the spectrum as broad, diffuse absorption bands (Figure 9.7).[10]

Despite the differences between Rayleigh scattering and Chappuis absorption, the hues that result are remarkably similar. Only experienced observers will perceive the difference. That is why Hulburt marveled at this coincidence, which had hidden the influence of ozone on the colors of the sky for so long. In 1956, when being awarded the Optical Society of America's Frederic Ives medal for his research, Hulburt wrote:

> The unsuspecting observer lying on his back and looking upward
> at the clear sky during sunset sees only that the overhead sky,
> which was blue before sunset, remains the same luminous blue
> color during sunset and throughout the darkening period of twi-
> light. He is not aware that in order to produce this apparently
> simple and satisfactory result Nature has dipped quite freely into
> her best bag of optical tricks.[11]

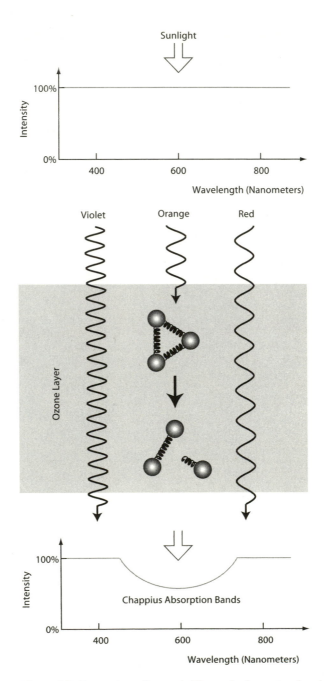

Figure 9.7 Formation of ozone's Chappuis absorption bands.

IN THE SHADOW OF EARTH

Around the time of Hulburt's discovery, the French meteorologist Jean Dubois noticed the influence of ozone in another twilight color phenomenon. Many people are dazzled by the warm colors of the western sky after the sun has set, but overlook a fascinating spectacle taking place behind their backs: the ascent of the Earth shadow. A few minutes after the sun has set in the west, the Earth shadow becomes visible in the east as a gray-blue arch over the horizon. It approaches the horizon in the north and the south, and reaches its highest point in the east (Color Plate 23). Above it, an arch of pink-colored light is seen. This is the light from the western twilight sky scattered back by the air molecules. As time progresses, orange and reddish colors in the west shrink to a bright segment, often called the twilight arch. Gradually the colors begin to pale and the light decreases in intensity. As the sun sinks lower, the backscattered light in the antitwilight arch pales, and the Earth shadow, which has risen in the meantime, dissolves without any distinct upper boundaries into the deep blue zenith of the sky. The last traces of the Earth shadow disappear about half an hour after sunset. These phenomena are visible in reverse order before sunrise; in other words, before the sun rises in the east, the Earth shadow will descend in the west.[12]

Sometimes the earth shadow is confused with the layer of haze near the horizon, but it is possible to tell them apart. In contrast to the earth shadow, the haze is already visible before sunset, and it lacks the former's characteristic gray-blue hue. The Earth shadow is visible only when the view of the eastern horizon is free of obstructions.

It is remarkable that the earth shadow was discovered fairly recently. In fact, even though the colors of twilight have long had a particular appeal to humans, occasionally even achieving the status of a deity (like Eos, the Greek goddess of dawn), no detailed descriptions were undertaken until rather late. The first on record is by the priest Johann Caspar Funck in southern Germany in 1716. Although he mentioned the pinkish glow of the antitwilight, he overlooked the Earth shadow. The Frenchman Jacques le Mairan was the first to describe it as the "dark

segment" in 1754. Three decades later, the Earth shadow was identified
by Horace Bénédict de Saussure, the inventor of the cyanometer (see
Chapter 5), during his ascent of Mont Blanc in 1787. Only since the
mid-nineteenth century have observations of the Earth shadow, using
this name, been regularly reported. At that time forays into the Alps
were becoming increasingly popular, so that many high mountains were
climbed for the first time. Since twilight phenomena are best seen from
higher elevations, the new interest in them is understandable. In 1868
even John Tyndall, who at the time was beginning his studies on turbid
media (see Chapter 6), made it to the top of the Matterhorn near Zermatt, Switzerland. He recorded the twilight phenomena on his way
down.

Interest in the Earth shadow was literally overshadowed by interest in
the alpenglow phenomenon. Just after the sun sets, the mountains opposite seem to radiate a pinkish glow, which seems to get brighter as the
light of the surrounding landscape fades. In Switzerland, Austria, and
Germany, this is known as *Alpenglühen* (alpenglow). By the 1860s there
was no doubt that it is a reflection of light from the setting sun that
gradually changes into a reflection of light from the twilight sky. No detailed studies of the Earth shadow were done until Jean Dubois began
his work in the 1940s. For three years his workplace was on yet another
mountain, the observatory on Pic du Midi in the Pyrenees, near the
border between France and Spain. This was an ideal place for his studies,
because it is particularly easy to see the Earth shadow from mountains,
where you are above the layer of haze on Earth's surface.

As its name implies, the Earth shadow was long considered to be the
shadow of the curved edge of Earth, which the setting (or rising) sun
was thought to project onto the opposite twilight sky. Arguments in favor of this view are its gradual ascent as the sun descends, as well as the
arch that spreads from the northern sky to the southern sky. Dubois
proved that this explanation was too simple. Intrigued by the Earth
shadow's unique gray-blue color, he investigated its spectrum. His investigation left no doubt: this color is due to the Chappuis absorption
bands of ozone. Just like the blue color seen in the zenith during twi-

Figure 9.8 Effect of the ozone layer on the colors of the sky at twilight.

light, it becomes visible because of the long path of sunlight through the atmosphere, which intensifies the optical effect of the ozone layer (Figure 9.8). Thus, when the weather is clear, we see the colored shadow of ozone in the eastern sky after sunset.[13]

Dubois also realized that our planet's curved edge could not explain why the shadow has a perceptible upper boundary. As mentioned earlier, the boundary is best detected just after sunset, and becomes increasingly diffuse as evening sets in. This observation is partly explained by the near coincidence of the shadow in the eastern sky with the line of sight of an observer looking east. Nevertheless, we cannot say that this is actually the shadow of the curved edge of Earth, since the boundary is much blurred by the effects of multiple scattering. In fact, it is the oblique incidence of the sun's rays along the ozone layer that makes the Earth shadow visible. If it were not for the gray-blue color, the shadow as such would hardly be noticed. Perhaps we should actually refer to it as the colored shadow of the ozone layer rather than the Earth shadow.

SEEING FAINT LIGHT

Rayleigh scattering and the absorption of light by ozone are not the only factors affecting our view of the twilight sky. It is also influenced by the way our own eyes adapt to low light levels. Our color perception changes during the transitions from day to twilight and from twilight to dark night.

During the day we see primarily by means of cones. These are the most numerous of the light-sensitive cells in the retina of our eyes. The three types of cones are sensitive to red, green, or blue light. The cones for green light are the most sensitive. As the brightness of an environment decreases, like during twilight, our eyes switch from seeing with cones to seeing with rods. Although they are more light sensitive than the cones, rods can only distinguish between bright and dark. Rods are especially sensitive in the blue-green spectral range. Thus, the transition from seeing with cones to seeing with rods makes a blue surface in reduced light appear brighter than a red surface, even if both appear equally bright during the day. This phenomenon was discovered by the Czech physiologist Jan Purkinje in 1825.[14]

THE ANTARCTIC OZONE HOLE

Ironically, the stratospheric ozone layer is less known to the general public than the hole that appears in it at certain times. This phenomenon was discovered by Joseph Farman of the British Antarctic Survey in 1985. For almost three decades, Farman and his colleagues had regularly measured the total ozone in the atmosphere above Halley Bay, a British research station in Antarctica. When the time series was carefully studied for its long-term trend, a dramatic decline was noticed in total ozone concentrations during the Antarctic spring (September to November). By the late 1970s and early 1980s, these were found to drop to half the concentrations recorded in the same months during the late 1950s and early 1960s.[15] Soon this finding was confirmed by data from TOMS, the Total Ozone Mapping Spectrophotometer, an instrument flown on several NASA satellites since the 1970s. After Farman and his colleagues had published their findings, the TOMS scientists looked into their data, only to find that their measurements showed the same decline. They had missed the discovery because of attributing the surprisingly low ozone concentrations to instrument error. Once recognized as meaningful, the TOMS measurements revealed the area over which the seasonal ozone

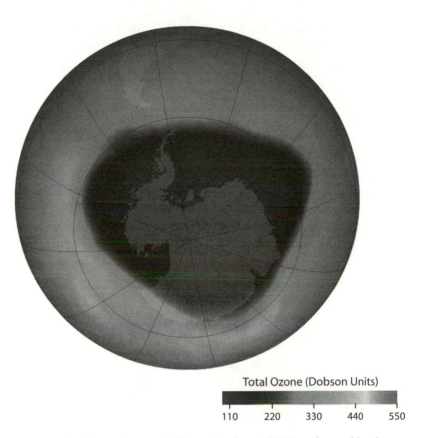

Total Ozone (Dobson Units)

110   220   330   440   550

Figure 9.9 The Antarctic ozone hole on October 1, 2005, as observed by the
Ozone Monitoring Instrument on board NASA's satellite *Aura*. The total amount
of ozone above each point on the earth's surface is indicated by the grayscale shading. In 2005, the ozone hole existed between early August and late October, with
a maximum in mid-September. Courtesy of NASA.

decline occurred. Since 1980 it has extended over the entire Antarctic
continent at its maximum (Figure 9.9).

Farman and his colleagues knew that man-made gases like halogens
(especially the chlorofluorocarbons, CFCs) and nitrous oxides (NO_x) in
the atmosphere could profoundly affect the chemistry of stratospheric

ozone, and these were soon identified as the culprit of the decline. The former are molecules containing chlorine, fluorine, and carbon that were originally introduced as cooling fluids in refrigerators and air conditioners, as well as spray-can propellants. Some of these substances are catalysts that decompose ozone. Their status as catalysts means that after destroying an ozone molecule, the CFC molecule emerges from a reaction unchanged, and ready to attack more ozone molecules. The well-known ozone hole over Antarctica is a result of declining ozone levels caused by these substances.

But why should there be an ozone hole over Antarctica, when most sources of CFCs used to be in the Northern Hemisphere? There are two reasons, both of them meteorological. First, observations have shown that within two to five years, all emitted CFCs become well intermixed throughout the atmosphere. Given that, on average, CFC molecules stay in the atmosphere for up to a century, they are distributed fairly homogeneously. Second, the seasonal cycle in Antarctica has a peculiarity. When the Antarctic winter commences in May, an enormous circular air current called the Antarctic vortex builds up, preventing the exchange of air with the subpolar atmosphere. For a period of more than three months, the Antarctic air is left to itself, while the temperature drops to abysmal depths. The air over the Antarctic continent is particularly dry, so the formation of clouds is inhibited—clouds of water, that is. Over Antarctica, winter temperatures can drop below $-78°C$ in the stratosphere, making the formation of clouds composed of nitric acid and water possible. In the absence of these so-called polar stratospheric clouds (PSCs), most chlorine and bromine is bound up in rather inert chemical compounds that are of little harm to the ozone molecules. Yet, when the PSCs have formed in the Antarctic winter, the surfaces of their ice crystals are the place where catalytic ozone depletion occurs. For these mechanisms to be effective, another ingredient is needed: energetic ultraviolet light that helps break up weakly bound molecules, much as in the undisturbed ozone chemistry described above. Since there is no sunshine in Antarctic winters, Antarctic spring is the time when all the ingredients necessary for ozone depletion are in place, and thus the ozone hole is most prominent in September and October. When the seasonal Antarctic vortex breaks down by November, part of the ozone-depleted air escapes the Antarctic, spills over to moderate latitudes, and may cause a brief 10 percent decline in the total ozone over New Zealand, Australia, and South America.

One might guess that there should be an ozone hole over the Arctic as well. Indeed, observations show that an Arctic vortex forms annually during the northern winter. However, this vortex is not as stable as its Antarctic counterpart; occasionally it allows warmer, ozone-rich air to flow in from the sub-Arctic. While PSCs do exist in the Arctic, the conditions are much less favorable for their formation. As a consequence, rather than an Arctic ozone hole there is a kind of seasonal ozone trough in the Arctic. This is not to say that ozone depletion is only a problem in the polar regions. Though less dramatic than the formation of the ozone hole, a general decline in stratospheric ozone has been documented even in moderate and equatorial latitudes.

In the mid-1980s, the world was shocked by the news about ozone depletion and its harmful consequences. Hardly any environmental problem has provoked international political action as rapid and forceful as in regard to banning CFCs. This is a remarkable achievement that culminated in the Montreal protocol of 1987, which ordered a 50 percent ban on the production of CFCs by 1998. The protocol was strengthened at a meeting in London in 1990, where CFCs were fully banned by 2000. Further amendments were made at conferences in Copenhagen and Beijing. Even though atmospheric levels of CFCs have started to decline, it may take half a century before the Antarctic ozone hole ceases to re-emerge. But success is not guaranteed. First of all, stratospheric chemistry is an extremely complicated business where scientists have been confronted with one puzzling enigma after another. In addition, banning CFCs may be a good thing for stratospheric ozone, but air traffic, predicted to increase significantly during the twenty-first century, may not. Its effects on ozone chemistry remain a subject of controversy.

If ozone is involved in coloring the twilight sky but is depleted seasonally over Antarctica, you may be wondering if any changes in the colors of the sky have been observed there. I had the same question and directly inquired with scientists at the Alfred Wegener Institute in Bremerhaven (Germany). Since 1980, the institute has been running the

Neumayer research station in Antarctica's Queen Maud Land. They said
no color effects had been noticed so far, at least not with the naked eye.
But then, nobody really seems to have been looking for it. Precise optical
investigations might detect a change in the color of the sky where ozone
is depleted. At the time of this writing, no such work has been done, and
thus the blue hour has not yet revealed all its fascinating secrets.[16]

CHAPTER 10

The Color of Life

To observers of the night sky, the star named HD209458 presents a fairly inconspicuous appearance. Situated in Pegasus, midway between the famous rectangle of bright stars and the neighboring constellation Delphinus, it can easily be seen with binoculars (Figure 10.1). Astronomers became interested in HD209458 in the 1990s when they realized that it closely resembles our sun if placed at a distance of 150 light years. It has just about the same mass, radius, and luminosity as our daytime star. In 1997 they found an even more compelling reason to watch this star closely. Spectroscopic observations revealed that it shows a characteristic wobble, apparently moving back and forth over a 3.5-day period. Astronomers unanimously concluded that this wobble must be due to a planetary companion of HD209458, with two-thirds the mass of Jupiter, or 200 times the mass of Earth. They named this extrasolar planet HD209458b.

Since 1995, when the first extrasolar planet was discovered orbiting the star 51 Pegasi, astronomers have been discovering such objects almost routinely. However, the planet orbiting HD209458 turned out to be in a special position. In 1999, observations by David Charbonneau of Harvard University and his colleagues showed that the star seems to drop in brightness by 1.5 percent for three hours every 3.5 days (Figure 10.2). The most plausible explanation was transits of the dark planet in front of the bright star. This meant that the planet must be a giant four times the diameter of Jupiter. The alignment of its orbital plane with our line of sight allowed a number of detailed observations that were impossible for any other extrasolar planet, and thus, in 2002, HD209458 hit the headlines. A group of astronomers around Charbonneau, now at the California Institute of Technology in Pasadena, had used the Hubble Space Telescope to discover a layer of sodium gas surrounding the planet HD209458b.

Figure 10.1 Since 1997 the star HD209458 (center) in the constellation Pegasus has been known to host at least one giant gaseous planet. The cross is caused by reflections in the telescope. Courtesy of Digital Sky Survey/National Geographic Society.

Shortly thereafter, a team of French astronomers headed by Alfred Vidal-Madjar of the Institut d'Astrophysique in Paris determined that there is an extended envelope of hydrogen gas as well. For the first time, a planet outside our solar system had been shown to possess an atmosphere.

This finding seems like a huge step toward discovering a second Earth, a planet comparable to our own, and one that may even harbor evolved life. Unfortunately, HD209458b and the other extrasolar planets discov-

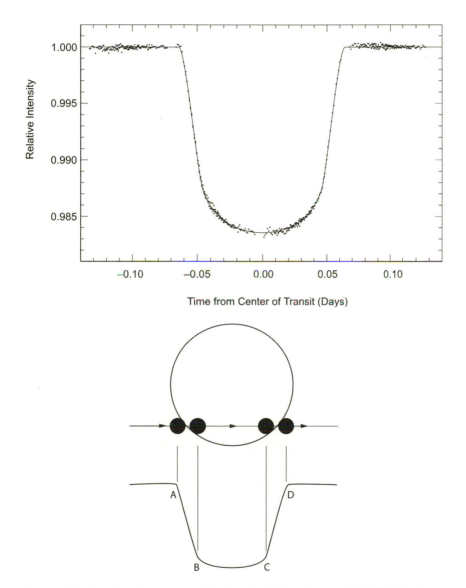

Figure 10.2 For three hours every 3.5 days, the light of the star HD209458 is diminished by 1.5 percent (above). This is interpreted as the visible effect of transits of a planet in front of its central star (below). The dark planet's transit in front of the bright star's disk begins at A, and a dip in the star's apparent brightness becomes noticeable. Between points B and C, the entire planetary disk is seen in front of the star. At point D the transit ends and the star resumes its pretransit brightness. Courtesy of Timothy Brown/American Astronomical Society.

ered so far cannot be counted on to make this astronomer's dream come true. Called "hot Jupiters," most of these objects are gas-rich planets orbiting their central stars in utter proximity. With atmospheric temperatures between 1,000 and 2,000 Kelvin, they are inhospitable to life. And what is more, at least in the case of HD209458b the atmosphere even seems to be vaporized; the planet drags its hydrogen envelope along like the tail of a comet.

Nevertheless, this is just the beginning of an exciting era, and it is only a matter of time until we find Earth-mass planets orbiting stars other than the sun. No astronomer doubts that such objects exist. But how can we tell whether or not they harbor life? If we were to receive a radio or TV broadcast, as unlikely as that seems, then this question would be settled in a dramatic way. However, we know today that more than three billion years passed from the origin of life on Earth to the invention of radio and television. If we wait for an extraterrestrial TV broadcast, we may overlook a great many inhabited worlds. Studying planetary atmospheres seems like a more promising way to find life on other worlds.

A first attempt to do so was made in 1990. The target was not some far-away exoplanet but our own cosmic home, Earth. On December 8 of that year, the NASA spacecraft *Galileo*, launched in October 1989, passed by our planet, taking a gravitational swingby on its way to Jupiter. The closest approach was 960 kilometers (600 miles) over the Caribbean Sea. Named after the Italian astronomer credited with the discovery of Jupiter's four largest moons in 1609, the spacecraft was equipped with a set of cameras and spectrometers to analyze electromagnetic radiation, including visible and infrared light.

As *Galileo* neared Earth, its instruments were turned toward our planet. They recorded data for analysis by a team led by the late Carl Sagan, a professor of planetary studies at Cornell University in upstate New York and renowned popularizer of astronomy. Even though the resolution was insufficient to make out man-made objects on Earth's surface, the team concluded that the measurements hinted at the presence of life; the atmospheric abundance of oxygen, ozone, water vapor, and methane being particularly noteworthy. The rationale was that the presence of life was necessary to maintain the chemical disequilibrium of Earth's atmospheric composition. Besides these "atmospheric biosignatures," spectral analysis of the planet's surface showed a sharp rise in emission at characteristic near-infrared wavelengths, a typical feature of plant foliage, and as such a "surface biosignature." And not only that, but there were also narrow-band, pulsed radio transmissions that Sagan and his team considered uniquely attributable to intelligence.

Now, a few years down the road, scientists are carrying on intense discussions about potential biosignatures in the spectra of distant planets. Preparations have started for the satellite missions *Darwin* and TPF (Terrestrial Planet Finder), which are meant to provide images and spectra of extrasolar planets sometime after 2015. There is hope that traces of life may be found in the spectra of these objects, but all the studies done so far suggest that this is going to be a very challenging task.

BLUE PLANETS

The findings of Sagan and his colleagues remind us that Earth is an oasis in the cosmos. Though we cannot rule out the possibility that life has arisen on other planets in the solar system, this is the only place where it has been able to continuously evolve over a long period of time. Earth is the only case we know of where life has played a formative role in planetary evolution. And not only is it unique as a habitat, but its apparent color is also special. It is often said that Earth is the only blue planet around, but this is not quite true; Uranus and Neptune look blue as well, although with a greenish tint (Color Plate 24). However, there are some crucial differences between these two planets and our Earth. Uranus and Neptune are huge balls of gas with dense atmospheres made up of light gases. They probably have only small solid cores, and they are far too cold for liquid water—a prerequisite for life as we know it. So it comes as no surprise that the blue color of their atmospheres is not caused by Rayleigh scattering but rather by the absorption of reddish-orange sunlight by methane gas, which is abundant on Uranus and Neptune. This means that Earth is the only "Rayleigh-blue" planet in the solar system. Is this exceptional color related to the presence of life on it?

Earth does in fact seem perfectly equipped to serve as a home to the life forms that we know of, and at the same time it has the bluest sky possible, at least as far as Rayleigh scattering is concerned. Meteorologists Craig Bohren and Alistair Fraser of Pennsylvania State University demonstrated this with a computer simulation (Figure 10.3). Assuming the atmosphere consists entirely of molecules that scatter sunlight according to Rayleigh's theory, their quantity present in the atmosphere turns out to be a near-perfect compromise between the saturation of the blue color and the brightness of the sky. While a tenfold increase in the amount of air would brighten up the sky, it would also dilute the saturation of the blue color, making it look paler. On the other hand, the color of the sky would be even more saturated than it is if the amount of air were reduced to one tenth, but the sky would then be so dark that we would hardly be able to make out the color.

Aside from its quantity, the composition of the air that surrounds Earth is also conducive to a blue sky. Oxygen gas has a threefold significance in this context: first, together with nitrogen its molecules make up 99 percent of the Rayleigh scattering particles. Second, the ozone layer— and thus the blue of the twilight sky—is derived from this gas. Finally, oxygen is the basis of the two most important atmospheric "cleansing agents": ozone (O_3) and hydroxyl radicals (OH). Because of their high reactivity, both of these attach themselves to foreign particles and separate them out of the air. This is important, because even a tiny quantity of aerosol particles will significantly decrease the saturation of the blue color.

If we compare the composition of Earth's atmosphere today with those of the two Earth-like planets, Venus and Mars, the first thing that stands out about the former is its high oxygen content (Table 10.1). It seems reasonable to attribute this to organic life on Earth, since this gas probably was as scarce in the atmosphere of the uninhabited young Earth as it is on the neighboring planets, which still lack an ozone layer to this day. The small percentage of carbon dioxide in Earth's present atmosphere is also noteworthy, since this gas is by far the predominant atmospheric component on Venus and Mars. In terms of surface temperatures, Earth seems to be the only habitable planet of the three. It is the only one

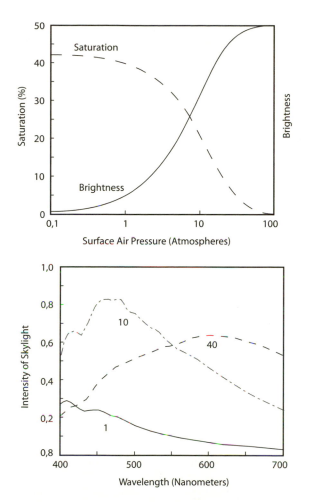

Figure 10.3 On Earth, the present quantity of atmospheric gases is especially conducive to the visibility of a Rayleigh-blue sky. Calculations performed by meteorologists Craig Bohren and Alistair Fraser have demonstrated this. The upper diagram shows the saturation of the blue color (in percent; dashed line) and the brightness of the sky (in arbitrary units; continuous) as a function of air pressure. The unit of air pressure used here is the current pressure of Earth's atmosphere at sea level. The diagram below shows the spectral intensity distribution of the sky's visible light in the present atmosphere (1; continuous line), as well as in hypothetical atmospheres with the same composition, but with ten times (dot-dashed) and forty times (dashed) the mass. Reproduced from Craig F. Bohren and Alistair B. Fraser, "Colors of the Sky," *Physics Teacher* 23 (1985), 271 (above), and Craig F. Bohren, "Atmospheric Optics," in *Encyclopedia of Applied Physics*, ed. G. L. Trigg, (Weinheim: VCH Verlagsgesellschaft, 1995), 12:411 (below).

TABLE 10.1

THE OBSERVED PROPERTIES OF THE ATMOSPHERES OF VENUS, EARTH, AND MARS AS
COMPARED WITH A (HYPOTHETICAL) LIFELESS EARTH

Property Evaluated	Venus	Earth Without Life (Hypothetical)	As It Is	Mars
Distance to sun (million of kilometers)	108	150	150	228
Radius (kilometers)	6,049	6,371	6,371	3,390
Solar irradiance (Watts/m^2)	2,631	1,367	1,367	589
Nitrogen content of atmosphere (% by vol.)	3.2	<2	78	2.7
Oxygen content of atmosphere (% by vol.)	Small traces	Small traces	21	Small traces
Carbon dioxide (% by vol.)	96	98	0.035	95
Surface pressure (atm)	90	60	1	0.0064
Greenhouse warming (°C)	+466	+270 ± 50	+33	+3
Average surface temperature (°C)	427	290 ± 50	15	−53

SOURCES: James E. Lovelock, *Gaia: A New Look at Life on Earth* (Oxford: Oxford University Press, 1979), 39; Bruce M. Jakosky, "Atmospheres of the Terrestrial Planets," In *The New Solar System*, ed. by J. Keally Beatty, Carolyn Collins Petersen, and Andrew Chaikin, 4th ed. (Cambridge, Mass.: Sky Publishing Corp. 1999), 176.

where water is found in a liquid state, whereas on Mars it is frozen and on Venus it occurs in gaseous form.

The young Earth probably did not have a blue sky. For one thing, the large number of carbon dioxide and water molecules must have diluted the saturation of the blue color significantly. Today we see this effect in the pale, bleached-out sky down near the horizon, which gives us an impression of how light is scattered by a quantity of air thirty-five times greater than that found in the direction of the zenith. The young Earth's atmosphere contained twice this many molecules, so that there probably was not much left of the Rayleigh blue. The lack of oxygen reduced the air's self-cleansing capacity, which in turn allowed the aerosol particles spewed out of the many volcanoes active on the young Earth to linger in the air for a long time. Since there was no ozone layer, there could not have been any "ozone's blue hour" at twilight. Back then, about four billion years ago, the sky was probably a great deal brighter and was either pale white or, due to aerosol and sulfur clouds, perhaps even yellowish in color. At any rate, it was still a long way from blue. The stations of its journey included the evolution of its atmosphere and the origin of life. If we want to understand how the blue sky color came about, we must travel back in time to the birth of our solar system.

HOW TO MAKE A PLANET HABITABLE

The sun and its planets formed about 4.6 billion years ago in a primeval nebula, a rotating disk of gas and dust. Today, we can infer the chemistry of this cloud from the present composition of the solar system. We learn that, among other ingredients, the solar system grew out of the remnants of one or more stars that had recently exploded as supernovae. This is the only way to explain the presence of gold and other heavy metals, as well as the high levels of radioactivity of the young planet Earth. Computer simulations suggest that the dust in this disk agglomerated into ever larger bodies, the so-called planetesimals, which collided with each other and, over the course of about 100 million years, formed the planets of the inner solar system: Mercury, Venus, Earth, and Mars. At that time, the predominantly gaseous planets of the outer solar system (Jupiter, Saturn, Uranus, and Neptune) were already radically different from these solid planets. Earth most closely resembled its neighbors Venus and Mars. Although it is possible that the three of them already had atmospheres at this stage, such atmospheres could hardly have survived up to the present;

every now and then, giant chunks of rock must have come crashing into them out of the primeval nebula. The energy released would have vaporized any atmosphere that had formed.

The heat generated by these collisions between planetesimals, together with its own gravity, caused several layers to form inside the young planet Earth, comparable to the layers of an onion. Heavy-metal compounds formed the core, around which the mantle and the crust settled. Earth's core, which was already in place 4.4 billion years ago, played a pivotal role in the history of the atmosphere, because radioactive decay within its depths provided the energy for volcanism, continental drift, and mountain formation. The volcanic gases built up the primordial atmosphere, a gaseous envelope vastly different from the present-day atmosphere. Its main components were water vapor (H_2O), carbon dioxide (CO_2), and molecular nitrogen (N_2), as well as traces of methane (CH_4), ammonia (NH_3), sulfur dioxide (SO_2), and hydrogen (H_2). At that time Earth must have been an inhospitable place, since large meteor impacts continued to occur, releasing enough energy to vaporize the young oceans.

About 3.8 billion years ago, the cosmic bombardment finally let up, and 300 million years later, or 3.5 billion years ago, the first organisms may have appeared. Warawoona Group in Western Australia is one place where palaeontologists have searched for their preserved remains. At this remote site in the outback, sedimentary rocks have remained virtually unchanged for most of Earth's history—an exceptional circumstance, considering that the face of our planet has been altered many times over by the forces of plate tectonics, volcanism, and weather. Of course, the rocks of Warawoona are weathered as well, and their red colors attest to the oxidation of the iron contained in them. It may seem overly optimistic to search for tiny traces of the first life forms at a site like this. However, it was here that scientists found wavy laminations in the sedimentary rocks that bear a stunning resemblance to stromatolites, formations of microbial mats known from presently existing shallow lagoons in the Caribbean and the Australian Great Barrier Reef. In the early 1990s, the discovery of these structures by Don Lowe of Stanford University and his colleagues caused a stir among geologists. A decade later, however, Lowe retreated from his

original position. While the age determination of the rock formations remained sound, he had to admit that the structures may have been caused
by inorganic chemical processes. The debate rages on to this day.

There is another find from Warawoona that has been interpreted as a
trace of early life. Cherts found in the same formation, when cut in thin
slices and viewed under a microscope, reveal microstructures that look
like fossil bacteria to some paleontologists. However, critics claim that
these structures are merely chains of crystals formed in hydrothermal
veins. Finds from old rock formations in South Africa and Greenland
show similar features. As controversial as these specimens are, even more
uncertainty surrounds the interpretation of finds from Akilia Island off
southwestern Greenland, which could date the advent of life to as much
as three hundred million years earlier.

The scientific disputes notwithstanding, evidence seems to be
mounting to the effect that life seized the first opportunity to establish
itself here on Earth after our planet's stormy beginnings. To some astro-
biologists, this suggests that life could be less of an exception and more
of a widespread phenomenon in the universe.

The earliest algae were capable of photosynthesis, using the energy of
sunlight to split water and carbon dioxide molecules, thus producing
carbohydrates. Through this process, molecular oxygen (O_2) is freed and
released into the atmosphere, while carbon is separated out of the air. By
extracting carbon dioxide from the air dissolved in the water and freeing
molecular oxygen, the blue-green algae began to steadily alter the composition of the atmosphere—provided that carbon was deposited in sediments. At first, the amount of oxygen freed in this way did not exceed
what could be used up by other geochemical processes. But by about
two billion years ago, the atmospheric oxygen concentration was already
at 1 percent of the present value and climbing. About 700 million years
ago there was 10 percent as much oxygen as today, and about 400 million years ago the oxygen in the air reached its present fraction of 21
percent by volume (Figure 10.4).

This increased oxygen concentration sparked a revolution in the history of the planet Earth. Up until then, the atmosphere had been reducing, which means that the oxygen freed by organisms was quickly

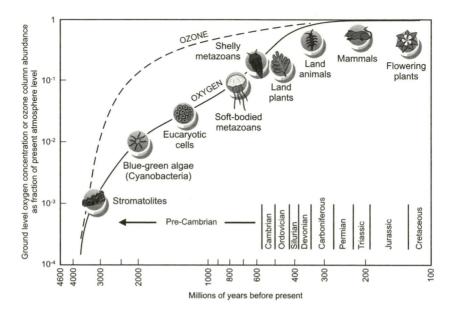

Figure 10.4 In Earth's present atmosphere, oxygen is the most prevalent gas after nitrogen. It has accrued in such high concentrations because of plant photosynthesis and the removal of carbohydates from the air by geological processes over the past three billion years. After Richard P. Wayne, *Chemistry of Atmospheres*, 3rd ed. (Oxford: Oxford University Press, 2000), 672.

used up in chemical reactions. There were compounds in the sea and air which could react with oxygen when they came in contact with it, thus becoming oxidized. However, about 600 million years ago, the rate at which the growing bacterial populations were producing oxygen surpassed the consumption of the gas in chemical reactions. As a result, first the atmosphere and soon the ocean as well became oxidized, as many chemical substances that had previously been dissolved in air and water combined with oxygen. Only in oxygen-free niches could these substances continue to exist in their uncombined state. For the bacteria whose metabolism had set this process in motion, the outcome was a disaster—after all, they depended on an oxygen reducing environment. What resulted may have been the worst crisis in the history of life on Earth, as the previously dominant anaerobic (oxygen–adverse) organisms largely died out. For aerobic (oxygen-loving) organisms, on the other hand, a heyday was beginning. Within 200 million years, species that could use the oxygen contained in water and air emerged.

But life was not the only thing to undergo radical changes during this period; the color of the sky was affected as well. The chemical reactivity of the oxygen in the atmosphere must have resulted in large quantities of aerosols becoming oxidized and thereby eliminated from the air. Many of the larger aerosols, which had previously bleached out the color of the sky by acting as Mie scattering particles, could no longer dilute the blue of the Rayleigh-scattered light. The mass of the atmospheric gases, which was initially sixty times its present value, had decreased considerably in the meantime, so that multiple light scattering diminished in significance. At the same time, thanks to the increased supply of oxygen, the stratospheric ozone layer was able to form. This not only made the ozone's blue hour possible, but also offered protection against the sun's ultraviolet radiation. It was probably at this point that the sky turned blue. Most likely, the formation of the ozone layer and the blue sky immediately preceded the colonization of dry land by plants and animals.

While we can only speculate about exactly when the blue sky appeared, it is likely that life on earth exerted a formative influence on the color of the sky. There is no doubt that microbial life has had a strong impact on the present composition of Earth's atmosphere; practically all of the oxygen it contains is a product of photosynthesis. Perhaps the same can be said of the low concentration of carbon dioxide. Its concentration has dwindled to one thousandth of what it was in the primordial atmosphere. Even though green-leafed plants carry out most photosynthesis on land, they are not the source of an increase in atmospheric oxygen. This is because the oxygen they release in photosynthesis is balanced by the respiration of animals and the decay of plants. In contrast, marine photosynthesis by algae and plankton causes a net gain in oxygen and a burial of carbon in oceanic sediments.

Some experts think that inorganic processes could also remove atmospheric carbon dioxide and thus reduce the number of scatterers. According to this view, high levels of atmospheric carbon dioxide must have set a greenhouse effect in motion, and the resulting temperature increase caused large quantities of water to evaporate. Then, precipitation washed the carbon dioxide out of the air. At Earth's surface, this gas would have been drained out of the air through the formation of carbonate rocks.

With or without the impact of life today Earth's carbon is to a large extent bound up in rock, while on Venus the majority of the carbon is located in the atmosphere in the form of the greenhouse gas carbon dioxide. The nitrogen contained in our atmosphere, which at present still takes up 78 percent of the air's volume, was not affected by biological processes. Rather, it stems mostly from the volcanic gases emitted over the last 4.5 billion years. In contrast, argon—the third most prevalent gas in the atmosphere by volume—accumulated there as a product of radioactive decay.

In this brief sketch of how Earth's atmosphere evolved, we have left out one baffling enigma. Since the first blue-green algae appeared on the scene 3.5 billion years ago, Earth appears to have been continuously inhabited. Moreover, up until 400 million years ago the sea was the most important—if not the only—habitat on the planet. Living in water requires a temperature above the freezing point. The highest possible temperature for this habitat is determined by the boiling point of water. Cells that contain protein can only exist at temperatures up to about 60°C (140°F), while some bacteria in geysers and deep-sea volcanic vents called "black smokers" thrive in water or steam ranging up to 125°C (257°F). Accordingly, the surface temperature of Earth can only have fluctuated between 0°C and 125°C.

It was geologists and biologists who proposed these narrow constraints on the environmental conditions for life to evolve on Earth. In order to be credible, they must be consistent with the findings of other scientists—of astronomers, for instance. During the last few decades, astronomers have made great progress with computer models of the structure and evolution of stars, and they believe that they understand the sun particularly well. The sun is an average star, and many predictions about its structure and internal properties have been confirmed with flying colors. For example, the estimated temperature at the center of the sun,

15 million Kelvin, has been confirmed by observing a characteristic pattern of sound waves on the solar surface that extends throughout the sun's interior and corresponds to its thermal structure. One feature of the standard model of solar evolution is that its energy output should have increased by 30 percent during the first three billion years of Earth's evolution. But this adds up to one big paradox; the very constant temperature of the biosphere would seem to contradict the increase in solar radiation, which should have effected a considerable temperature increase on Earth.

EARTH IN A GREENHOUSE

The solution to this paradox of the "faint young sun" involves a process that was first noticed by Jean Baptiste Joseph Fourier in 1824.[1] This French mathematician claimed that Earth's atmosphere was comparable to a greenhouse, in that it lets the sun's light through but holds in the heat radiation reflected off the surface of Earth. This "greenhouse effect," he said, had warmed up the atmosphere and the surface of our planet. Fourier thought this was the mechanism that made life on Earth possible. Without it, Earth would be a frozen, lifeless ball of rock hurtling through space.

Researchers began to understand the cause of this hypothetical process when they looked at how infrared radiation affects various gases. Forty years later John Tyndall realized that nitrogen and oxygen gas, which together take up 99 percent of the air's volume, have no influence on the greenhouse effect, since neither of them affects the propagation of heat radiation. This is not the case with water vapor and carbon dioxide, about which Tyndall writes:

> But the aqueous vapour, which exercises such a destructive action on the obscure [i. e. invisible heat] rays, is comparatively transparent to the rays of light. Hence the differential action, as regards the heat coming from the sun to the earth and that radiated from the earth into space is vastly augmented by the aqueous

vapour of the atmosphere. . . . Similar remarks would apply to the
carbonic acid diffused through the air, while an almost inapprecia-
ble admixture of any of the hydrocarbon vapours would produce
great effects on the terrestrial rays and produce corresponding
changes of climate.[2]

Water vapor and carbon dioxide thus have the property of transmitting
visible light while absorbing infrared or heat radiation. All gases that have
this property are called greenhouse gases. As Fourier and Tyndall realized, they play an important role in regulating the climate. Most of the
solar energy that reaches our planet is in the form of visible light, which
heats up the surface and is radiated back partly as infrared radiation
(heat). Greenhouse gases in the air prevent this heat radiation from dissi-
pating out into space, so that it stays in the atmosphere instead. As a result, the atmosphere and Earth's surface are heated up.

Thirty years later, and with Tyndall's findings in mind, the Swedish
chemist Svante Arrhenius performed an involved calculation and came
to the conclusion that a 50 percent drop in the atmospheric carbon
dioxide concentration would cause the atmosphere to cool by 4 to 5°C
(7 to 9°F). Arrhenius deduced from this that fluctuations in the concentration of this relatively rare atmospheric gas could have triggered the
ice ages and warm periods of the past millennia. Yet he did not know
whether the proportion of carbon dioxide in the atmosphere could really fluctuate widely enough to trigger an ice age.

Today we know that the naturally occurring quantities of water vapor
and carbon dioxide in the atmosphere raised the average surface temperature of Earth from −18°C (0°F) to +15°C (59°F), thus making our
planet habitable. This natural greenhouse effect is a stabilizing part of the
climate that has protected life on Earth from looming crises for over
three billion years. Looking back on the paradox of the "faint young
sun," the greenhouse effect acted as a midwife for life on Earth—and at
the same time, it is part of a complex interaction between the atmosphere and the upper layers of the Earth's crust. If the reduced incoming
solar energy in the past had led to a lower surface temperature here on
Earth, this would have diminished the (temperature-dependent) capacity

of weathering rocks to absorb carbon dioxide, thus removing it from the atmosphere. At the same time, this greenhouse gas continued to be emitted by volcanoes, and it must have accumulated in the air. The result was an amplified greenhouse effect, which raised the temperature of the atmosphere. However, this in turn increased the rate at which the rock could bind carbon dioxide, so that the amount of the gas remaining in the air stabilized at a certain level. The greenhouse effect of carbon dioxide thus appears to be a mechanism that regulates the Earth's surface temperature, holding it constant within a certain range even during fluctuations of solar energy output.

Up until about 200 years ago, the Earth's atmosphere was in relative equilibrium. The production of oxygen by organisms matched its consumption exactly, while the greenhouse effect stabilized the temperature. Of course climate fluctuations often occurred, for example during the ice ages and as a result of occasional meteorite impacts, one of which may have led to the extinction of the dinosaurs 65 million years ago. But on the whole, life on Earth was relatively secure. And it contributed to its own security by producing oxygen, for example. This cannot belie the chemical instability intrinsic to the Earth's atmosphere. Assuming plant photosynthesis were to come to an abrupt halt, the present supply of oxygen would be used up by organisms and chemical reactions within just four million years. Compared with the two billion years it took to build up the present supply of oxygen, this is a decidedly short time. Because of the important properties of this gas, such a drop in the oxygen levels would probably mean the end of the blue sky as well.

REDISCOVERING THE BIOSPHERE

An alternative solution for the paradox of the "faint young sun" was proposed by the English scientist James Lovelock at the beginning of the 1970s. He claims in his Gaia hypothesis that living beings not only control the oxygen content of the air, they also regulate the temperature of the atmosphere. Lovelock points out that our Earth seems optimally equipped to host life. The average temperature on its surface ($+13°C/+55°F$) and the atmospheric oxygen concentration (21 percent) are at the ideal levels for higher organisms. Even a slightly larger percentage of oxygen would increase the frequency of forest fires considerably; on the other hand, if it were smaller, then many species would not be able to survive. Lovelock suggests that these limits are not coincidental but rather are determined by the planet's organisms. He suspects that the living beings on Earth regulate the temperature and composition of both biosphere and atmosphere in order to maintain ideal living conditions for themselves. In other words, they are not just exploiters, but creators of these conditions. Lovelock sees Earth as a living organism, which he has named Gaia after the Greek Earth goddess.

Though this theory made Lovelock a controversial figure in environmental science, his idea does have its merits. The ideal oxygen concentration of today's atmosphere fits the model, as does his solution for the paradox of the "faint young sun." Lovelock claims that plants have been able to stabilize the temperature of the air by continually reacting to the temperature of their surroundings and the intensity of sunlight. He illustrates this claim with the "Daisyworld" model, a hypothetical Earth inhabited by two species of daisies. Dark daisies absorb solar energy and contribute to warming up the atmosphere, while light daisies reflect sunlight, thus cooling Earth down. Daisyworld would compensate for a long-term rise in incoming solar energy with a relative increase in the population of white daisies, up to the point at which all the daisies were white and the temperature could no longer be contained by these means.

But this model does not constitute proof of the Gaia hypothesis. Its critics mainly object to Lovelock's claim of goal-oriented adaptation of the biosphere and atmosphere. Wouldn't this ascribe consciousness to the oxygen-producing bacteria? Lovelock defends himself with the argument that regulation can also take place unconsciously. However, one serious challenge to his theory remains in the indisputed transition, about 600 million years ago, from a reducing atmosphere on Earth to the oxidizing atmosphere of today. This transformation was caused by the very bacteria who would later be poisoned by the oxygen they were producing. As a result of their photosynthesis, these bacteria nearly destroyed their own habitat, which contradicts the Gaia hypothesis. Though it may seem questionable that the Gaia hypothesis can be verified, Lovelock has still rendered the valuable service of opening many people's eyes to the complex interactions that exist between the atmosphere and the biosphere on "Spaceship Earth."

It turns out that the dynamic proposed by Lovelock is not entirely novel. Rather, it is reminiscent of ideas developed in the early twentieth century by Ukrainian-born geologist Vladimir Ivanovich Vernadsky. A professor at Moscow University in 1916, Vernadsky began to think about how living matter might transform solar energy and thus become a force in transforming the planet. At the time, it was known that coral reefs, fossil fuel deposits and carbonate rocks pointed to the importance of life in shaping Earth. But to Vernadsky, this insight was not far-reaching enough; he wanted to understand the importance of "the whole organic world in the general scheme of chemical reactions on the Earth."[3] After surviving the turmoil of World War I, the Bolshevik revolution and its aftermath, Vernadsky went to lecture at Sorbonne University in Paris in 1922. Leaving Paris for Prague in late 1925, he began writing his book *Biosfera* (*Biosphere*). It was published in mid-1926. To Vernadsky, the biosphere is the zone where life has a transforming effect on the environment, extending from high up in the ozone layer down to the bottom of the ocean and into the depths of Earth. He is fascinated by the interaction of sunlight, life, and the upper layers of Earth's crust in transforming our planet:

> The radiations that pour upon the Earth cause the biosphere to take on properties unknown to lifeless planetary surfaces, and thus transform[s] the face of the Earth. Activated by radiation, the matter of the biosphere collects and redistributes solar energy, and converts it ultimately into free energy capable of doing work on Earth.
>
> The outer layer of the Earth must, therefore, not be considered as a region of matter alone, but also as a region of energy and a source of transformation of the planet. To a great extent, exogenous cosmic forces shape the face of the Earth, and as a result, the biosphere differs historically from other parts of the planet. The biosphere plays an extraordinary planetary role.[4]

While Vernadsky thus understands the transformatory power of life on
Earth, he is also aware of the precarious balance needed to sustain it.

CHANGING THE AIR, HEATING THE PLANET

Svante Arrhenius pointed out that since the beginning of the Industrial
Revolution, the burning of fossil fuels has released large amounts of carbon dioxide into the atmosphere. Extending his calculations, he determined that doubling the atmospheric carbon dioxide concentration
would raise Earth's surface temperature by 5 to 6°C (9 to 11°F). He did
not see this as a problem. In his book *World in the Making* (1906), he
writes optimistically:

> By the influence of the increasing percentage of carbonic acid
> [carbon dioxide] in the atmosphere, we may hope to enjoy ages
> with more equable and better climates, especially as regards the
> colder regions of the Earth, ages when the Earth will bring forth
> much more abundant crops than at present, for the benefit of rap-
> idly propagating humankind.[5]

Perhaps Arrhenius' outlook was influenced by life in frigid Scandinavia.
Thirty years later, Guy Stewart Callendar, a steam technologist working
for the British Electrical and Allied Industries Research Association,
studied data from two hundred weather stations and concluded that the
air temperature on the surface of Earth had already risen by about
0.28°C from 1880 to the 1930s—which he considered the beginning of
the artificially intensified greenhouse effect.[6] Like Arrhenius, Callendar
thought this was a good thing, since a warming of the climate in Europe
would result in the expansion of the arable land area and ward off the
threat of a new ice age. As late as the 1980s, Mikhail Budyko and his colleagues from the Russian State Hydrological Institute in St. Petersburg
took the same stance, citing advantages of a temperature increase for (at
that time) Soviet agriculture.

The fact that large amounts of carbon have ended up in the atmosphere since the Industrial Revolution, and that humans are thus changing
the chemical composition of the atmosphere, has been indisputed since

Arrhenius' time. Yet some doubt remained as to Arrhenius' and Callendar's assumption that the carbon dioxide released by the burning of fossil fuels did in fact accumulate in the atmosphere. It has been known since the late nineteenth century that more carbon dioxide is dissolved in the water of the oceans than in the atmosphere. Thus it was possible to suppose that the carbon dioxide humans were releasing into the atmosphere would be absorbed by the oceans. But this was by no means a proven scientific fact, and, moreover, no one knew whether the carbon dioxide contained in the oceans had already reached the saturation point. In 1957, oceanographers Roger Revelle and Hans Suess called on researchers to perform measurements in order to answer this question:

> Human beings are now carrying out a large scale geophysical experiment of a kind that could not have happened in the past nor be reproduced in the future. Within a few centuries we are returning to the atmosphere and the oceans the concentrated carbon stored in sedimentary rocks over hundreds of millions of years. This experiment, if adequately documented, may yield a far-reaching insight into the processes determining weather and climate. It therefore becomes of prime importance to attempt to determine the way in which carbon dioxide is partitioned between the atmosphere, the oceans, the biosphere and the lithosphere.[7]

Revelle had been the director of the Scripps Institution of Oceanography in San Diego since 1951. From the beginning of his research career, he had been studying the chemistry of seawater, finding it to be characterized by complex, often counterintuitive processes. During the 1950s, estimates of how long carbon dioxide molecules remain in the atmosphere ranged from sixteen years to several thousand years. When Revelle wrote his article with Suess, he had already found a man who could measure the exact amount of carbon dioxide in the atmosphere and perhaps even detect a possible increase. That man was Charles Keeling, a postdoctoral researcher in geochemistry at the California Institute of Technology in Pasadena. In the spring of 1955 Keeling had succeeded in building an accurate manometer, a instrument that could detect small concentrations of gases. According to initial measurements carried out on the roof of the chemistry department, the carbon dioxide concentration was 315 parts per million. Keeling repeated the measurements at different times of day and in different weather conditions, and the carbon dioxide concentration always amounted to 315 parts per million. Pasadena is a suburb of Los Angeles, so it seemed quite possible that this value was influenced by the nearby metropolis. So Keeling took to the road, traveling into the remote Yosemite National Park. There, far away from any industrial facilities, he also found a carbon dioxide concentration of 315 ppm—an intimation of how thoroughly intermixed the atmosphere is.

When Revelle brought Keeling to the Scripps Institution in San Diego, he wanted him to document how this concentration was going to change. If it should remain stable, then that would point to the ability of the oceans and biosphere to quickly bind excess carbon dioxide. An increase, on the other hand, could indicate either a slow absorption rate or even a limited absorption capacity of the oceans. The prospects for undertaking a comprehensive series of measurements were good at the time, because the International Geophysical Year was about to commence. This was a worldwide research program in which scientists from sixty-seven countries performed coordinated observations of the atmosphere and the oceans, spanning from pole to pole. Part of the United States' contribution was the construction of an atmospheric observatory near the summit of Mauna Loa, a 13,680-foot volcano in Hawaii. Keeling wanted to install an improved manometer there to measure the carbon dioxide concentration far from sources of carbon dioxide emissions. In the spring of 1958, such a device was brought to the station.

The first measurements of the carbon dioxide concentration yielded a value of 314 ppm, which corresponded approximately with the measurements from California. But by the end of the year, the concentration rose to 318 ppm—then dropped back down to 314 the next spring. The following year (1959), this pattern repeated itself. The atmospheric carbon dioxide concentration rose significantly at the end of the year, only to drop off in the spring. Keeling and his colleagues realized that this pattern reflected the yearly carbon cycle in the biosphere—the shifts coincide mainly with the seasons of the northern hemisphere. At the be-

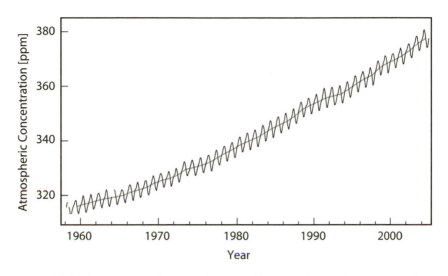

Figure 10.5 The concentration of carbon dioxide gas in the air (in parts per million), as measured at Hawaii's Mauna Loa Observatory. A continuation of measurements first performed by Charles Keeling in 1958, this curve is often called the Keeling curve. Courtesy of NASA/NOAA.

ginning of the growing season in spring, Earth's vegetation draws carbon dioxide out of the air, then gives it back at the end of the growing season in fall.

Keeling and his colleagues also recognized that there was another trend layered on top of this seasonal pattern: a continual increase in the average carbon dioxide concentration. When the measurement series had first started, the concentration was already well above the 290 ppm measured in the early 1870s. Since then, the carbon dioxide concentration has risen by about 2 percent per year, and even 3 percent during the last few years. This rising trend, when superimposed on the seasonal variation, produces a characteristic climbing curve that has long been known as the Keeling curve (Figure 10.5). It is probably the most well-known document in the debate on global warming. The carbon dioxide concentration amounted to about 326 ppm in 1970, about 338 ppm in 1980, about 354 ppm in 1990, and about 369 ppm in 2000.

While the measurements taken by Keeling and his successors at the

Mauna Loa Observatory document recent developments in the atmospheric carbon dioxide concentration, studies of ice cores from Greenland and Antarctica make it possible to trace past developments in carbon dioxide concentrations. This is done by studying air bubbles that have become trapped in the annual snow accumulations. In the mid-1990s, a French-Russian team drilled a core 3,300 meters down into the Antarctic inland ice sheet near the Vostok station. The core contains air bubbles from the past 420,000 years. Over this long period of time, the atmospheric carbon dioxide concentration was never as high as it has been since World War II.[8] Furthermore, climatologists have learned to use so-called proxy data, such as the isotope ratios contained in the ice, to infer the temperature of the air through which the snow fell before landing on the ice. These data show that, over the entire period in question, there was a close relationship between the atmospheric carbon dioxide concentration and the air temperature. Around the year 2000, annual anthropogenic emissions of carbon were at 6 to 7 billion tons. And climbing.

Many measurements have provided unmistakable proof of global warming. But since the 1980s, public discourse has returned again and again to the question of whether it is becoming detectable to the average person. It is true that the climate change cannot be pinned down just on the basis of individual, extreme weather events, such as unusually hot summers or droughts. But we know that the average surface temperature of Earth rose by 0.6°C (1°F) in the twentieth century, that the 1990s were probably the warmest decade in the past millenium, that the growing season in the temperate zones has become longer, and that the thickness of Arctic sea ice in summer has decreased by 40 percent within the last four decades.[9]

Every five years since 1990, the Intergovernmental Panel on Climate Change (IPCC), the leading brain trust on climatology, has published a status report documenting the state of the art in measurement data and climate models. The 2001 report introduces different scenarios of climate development that predict a temperature increase of 1.4 to 5.8°C (2.5 to 10.4°F) from now until the year 2100. The predictions depend on assumptions about future consumption of fossil fuels, as well as the transition to alternative energy sources. Initial calculations for the next IPCC report have already been published, and they forecast a temperature increase of 4.0 to 4.5°C (7.2 to 8.1°F) from now until the year 2100. These figures are based on the assumption that we will not manage to reduce carbon dioxide emissions significantly. Meanwhile, ecologists estimate that a temperature increase of more than 2°C (3.6°F) within one century will do serious harm to Earth's ecosystems, not to mention the intrinsic dangers of an escalating, self-compounding climate change. With all this in mind, the fact that the warming trend will not have an immediate effect on the color of the sky can be of little comfort.

LOOKING AT OUR COSMIC NEIGHBORHOOD, AGAIN

The picture drawn here makes it worthwhile to take another look at Table 10.1, the comparison of the atmospheres of Venus, Earth, and Mars. The initial, primordial atmospheres of these planets appears to have been rather similar and consisted of volcanic gases. The fact that these three atmospheres have evolved so differently seems to be due to differences in mass and incoming solar radiation. As we have seen above, the geological record on Earth shows that for most of its history liquid water could exist on its surface—a necessary ingredient for the evolution of life as we know it. A series of feedback mechanisms acting on different timescales have helped to sustain a habitable environment. On Venus and Mars such mechanisms must have failed altogether or have ceased in the course of time.

While Venus has about the same volume and mass as Earth, due to its closer proximity to the sun it receives twice as much solar energy as our planet. It seems that this started Venus off on a completely different evolutionary path. From the time when its crust was solidifying, there was probably less water becoming trapped in rock; instead, it remained gaseous right from the beginning. The incoming solar radiation caused Venus' gaseous water suppy to evaporate and be lost to interplanetary space while its large amounts of carbon dioxide remained in the atmosphere. Without water vapor, the inorganic removal of this gas from the atmosphere through the formation of carbonate rocks failed. Venus

Figure 10.6 Dense global
cloud patterns in the at-
mosphere of Venus as seen
in ultraviolet light by
NASA's *Galileo* spacecraft.
Courtesy of NASA/Jet
Propulsion Laboratory.

remained in a heat trap, its surface temperature stabilizing at an inhospitable +427°C. Today the sky on this planet is considerably lighter than
that of Earth. It looks pale white and is occasionally darkened by clouds
of sulfuric acid (Figure 10.6).

The atmosphere of Mars evolved differently. There the incoming solar energy is only half as great as on Earth, and although the Martian atmosphere consists predominantly of the greenhouse gas carbon dioxide,
its mass was too small to effect a temperature increase. Today the median
temperature on the surface of Mars is −53°C (−63.4°F), and the water is
frozen solid in polar ice caps. But this was not always the case; large,
dried-up riverbeds bear witness to liquid water in its history. Mars is
home to several volcanoes, among them Olympus Mons, the largest
volcano in the solar system. However, all of these are long extinct. Martian volcanism probably ceased because of the planet's small body cooling rapidly, and thus the principal source of greenhouse gases was shut
down, leaving nothing to prevent the atmosphere's persistent cooling. As

Figure 10.7 Clouds over the rim of the crater Gusev on Mars in a photograph taken by the Mars Exploration Rover "Spirit" in November 2004. Courtesy of NASA.

such, Mars was unable to enter the carbon-silicon cycle, a feedback mechanism that stabilizes temperatures in the atmospheres of geologically active planets over long time spans. The bottom line is that a "geologically inactive planet is unlikely to be able to maintain a stable climate or hold onto its atmosphere for geologically long periods," as the geologist James Kasting and the atmospheric scientist David Catling conclude.[10] Due to the small mass of the Martian atmosphere, the sky there is considerably darker at present than on Earth. Winds often stir up dust that gives the Martian sky a reddish appearance, and occasionally even clouds of ice molecules can be seen in it (Figure 10.7).

The comparison of Earth, Mars, and Venus shows that the greenhouse effect can stabilize a climate, but only within certain limits. A small evolutionary difference can render a planet uninhabitable in the long term—and can incidentally lead to an atmosphere that does not shine with the blue color of Rayleigh-scattered light. Observing this blue in the atmospheres of distant planets may not guarantee their habitability, but at least on Earth it is a product of exceptional developments in which microbial life played a key role.

Epilogue

The most superfluous question is nevertheless: "Why?"

Dezsö Tandori[1]

We have come to the end of our journey through the history of the blue sky. Starting from the child's question, why is the sky blue, it has taken us from Greek philosophy to twentieth-century physics, and from the structure of the universe to the history of life and its impact on Earth's atmosphere.

The many approaches that have contributed to explaining the sky's color forbid us to single out any approach as the only useful one. Those who were hoping for a single, straightforward answer as to why the sky is blue may be disappointed. But those who are open to nature being comprehensible in a variety of ways may well be relieved.

In researching this story, I learned about Albert Einstein's admiration for the Austrian physicist-philosopher Ernst Mach. This fondness may seem curious in light of the fact that Mach happened to be a critic of the early views of molecular reality when Einstein was one of its main proponents. Yet in September 1910, after Mach had suffered a stroke, Einstein went to see him at his home in Vienna. Though paralyzed and suffering from impaired speech, Mach was ready to engage in a debate about the molecular hypothesis—and was persuaded by Einstein's arguments. Einstein kept emphasizing that it was Mach's persistent questioning of what others held to be fundamental truths that impressed him early on, and in a way, the theory of relativity itself is an assault on the seemingly unassailable mechanistic worldview prevalent in the nineteenth century. Mach died in 1916. Reminiscing over this man's influence on his own thought, Einstein wrote in an obituary:

> Concepts that have proven useful in ordering things can easily gain authority over us such that we forget their worldly origin and take them as immutably given. They are then rubber-stamped as "thought-necessities" and "given a priori," etc. Such errors often make the path of scientific progress impassable for a long time.[2]

If we look at the history of the sky's blue color in these terms, it becomes a story about letting go of preconceived notions about nature, letting go of "thought-necessities." At the same time, it is about being ready to embrace new insights that may contradict received wisdom.

From a cosmic perspective, the blue sky that we enjoy on Earth is an extraordinary phenomenon in every sense of the word. The story of how the sky became blue is a story of the transformability of our planet's atmosphere. And indeed, our atmosphere has become subject to changes that are accelerating and seem to progress as fast as our understanding of light, color, vision, and the air. We are now at a unique point in the history of mankind and the history of our planet. Within a brief period of time we are transforming what was hitherto regarded as "natural" and immutable. In this sense we have to give up yet another thought—necessity. Our impact on the environment is unprecedented in its magnitude and suddenness. We keep changing the atmosphere, with global warming and ozone depletion among the key processes that will determine the world our grandchildren inhabit. Contemplating the blue sky may be a first step toward understanding the atmosphere as a fragile and malleable realm that we must quickly learn to care for if we want to have a future on Earth, the best of all planets we know.

Estimating the Height of the Atmosphere from the Duration of Twilight

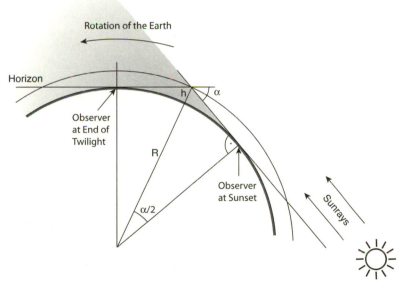

R   Radius of the Earth
α   Depression angle of Sun below the Horizon
h   Height of the Atmosphere

In his ingenious study of twilight, Abu'Abd Allah Muhammad ibn Muadh used simple geometry to estimate the height of the atmosphere (see Chapter 4). The illustration shows the round Earth with radius R surrounded by an atmosphere with thickness h. Owing to Earth's rotation, the illumination of the atmosphere changes over the course of a day. These changes are most dramatic during morning and evening twi-light. Evening twilight begins with the sun setting below the horizon. At this moment the entire atmosphere above an observer is directly illumi-nated by the sun. As the sun dips further below the horizon, progres-sively higher layers of the atmosphere fall into the shadow of the curved Earth. Eventually, the highest part of the atmosphere that is still illumi-nated by the sun sinks below the horizon. This moment marks the end of astronomical twilight and the beginning of night.

In the figure, α is the angle to which the sun has set below the hori-zon when the last light of evening twilight has disappeared in the west. It can be estimated from the sun's motion along its daily path. Abu'Abd Allah estimated that α amounted to 18 degrees—in perfect agreement with current measurements.

From the figure we can infer:

$$\cos\left(\frac{\alpha}{2}\right) = \frac{R}{R+h}$$

Solving for h yields:

$$h = R\,\frac{1 - \cos\left(\dfrac{\alpha}{2}\right)}{\cos\left(\dfrac{\alpha}{2}\right)}$$

If we plug in Earth's radius, $R = 6{,}371$ kilometers, and Abu'Abd Al-lah's $\alpha = 18$ degrees, we obtain $h = 79.5$ kilometers (49.4 miles). This is a reasonable estimate—and a valuable insight from a simple observation.

Blue Eyes as Turbid Media

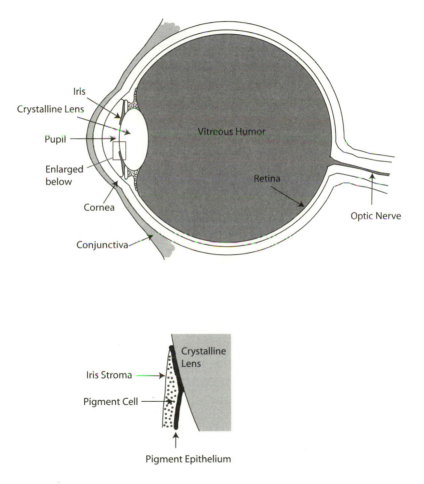

Soon after Ernst Wilhelm von Brücke had recognized the color changes
of chameleons as an optical effect of turbid media (see Chapter 6), his
friend Hermann von Helmholtz, a physicist and physiologist from Berlin,
made a related discovery. He found that the color of our eyes is another
optical effect of a turbid medium! If we examine the anatomy of the
eye, we find that its color is the color of the iris, a diaphragm that helps
our eyes adjust to the brightness of our surroundings. By expanding and
contracting, it determines the amount of light that passes through a
round opening into the pupil. The iris itself consists of two layers: at the
front is the stroma, a connective tissue interspersed with blood vessels,
and behind that the pigmented epithelium, which contains a large quantity of dark brown, nearly black pigment cells (figure). While the pigment
is abundant in the epithelium of all humans, its quantity in the stroma can
vary greatly from one person to the next. Blond-haired people typically
have a small concentration of pigment in their stroma, which functions
as a turbid medium in front of the dark-pigmented epithelium; their
blue eyes are thus an example of Goethe's basic phenomenon! Dark-
haired people, on the other hand, generally have more pigment in their
stroma, and the high concentration gives their eyes approximately the
same color as the pigment itself. Most newborns have a violet to blue-
green eye color that goes away over a period of months or years. This
shows that the pigmentation of the iris stroma increases after birth.

APPENDIX C

A Simple Derivation of the Inverse Fourth Power Law

John William Strutt, in 1870 the soon-to-be third Lord Rayleigh, wondered which physical variables might be involved in light scattering (see Chapter 7). For a first guess as to the wavelength-dependence of scattered light, he applied the "principle of similitude." This entailed combining these variables in a formula which describes the intensity of scattered light as a function of the wavelength of incident light, while keeping all the units of the equation balanced.

Let us relate, as Strutt did, the intensity I' of scattered light to the intensity I and the wavelength λ of an incoming light ray. Since this is an optical problem, the speed of light c might be of importance. The scattering particle's volume V might play a role as well. We can assume its linear dimension to be much smaller than the wavelength of the incident light, as the observed pattern of skylight polarization suggests. Moreover, the particles shall be spherical and randomly distributed in space. It is possible that the densities of the surrounding medium (D) and that of the particle (D') will enter into the formula. Finally, the distance r between the particle and an observer might also be relevant. The intensity of scattered light will decrease with increasing distance r.

Let us first compare the ratio of the amplitudes of the scattered (A') and the incoming light (A), or $\frac{A'}{A}$. Since A' and A have the same units, this fraction must be dimensionless. Thus, the densities D and D', as the only variables including the unit of mass, can only enter into the equation as a fraction $\frac{D'}{D}$ and cannot affect our dimensional considerations. Moreover, the speed of light cannot enter into the desired formula, since it is the only independent variable containing the unit of time (speed is the ratio of length over time). Therefore, the ratio $\frac{A'}{A}$ is governed by the three independent variables V, r, and λ.

If we assume larger particles to scatter light more effectively than small ones, we can anticipate that the ratio $\frac{A'}{A}$ will be proportional to the particle's volume V. Moreover, the energy of the incoming wave must equal the energy of the scattered waves. This is only possible if the intensity of scattered waves decreases with growing distance from the particle. Since the surface area of a sphere with radius r around the particle increases by the square of the radius, the intensity of scattered light must decrease inversely, that is, according to the formula $\frac{1}{r^2}$. This is a necessary condition for the total intensity of scattered light to equal the intensity of the primary wave. On the other hand, the intensity of a sinusoidal wave is proportional to the square of its amplitude. In reverse this means that $\frac{A'}{A}$ is proportional to $\frac{1}{r}$. Thus we have so far:

$$\frac{A'}{A} \approx \frac{V}{r}$$

Here, the dimensions have become unbalanced: while the left-hand side is dimensionless, the right-hand side has the unit of length squared. Note that our third independent variable, the wavelength λ, has the unit of length. Squaring it and placing it in the denominator restores the units:

$$\frac{A'}{A} \approx \frac{V}{r\lambda^2}$$

Eventually, since the intensity of a wave is proportional to the square of its amplitude, we obtain for the ratio of intensities of the scattered and the incoming wave:

$$\frac{I'}{I} \approx \left(\frac{A'}{A}\right)^2 \approx \frac{V^2}{r^2\lambda^4}$$

With this formula, Strutt had arrived at the desired result. If we interpret this formula as describing the light scattering off one single particle, V and r remain constant and the ratio of intensities only depends on the wavelength λ. The result is Rayleigh's inverse fourth power law of light scattering at small particles.

Atmospheric Extinction and Avogadro's Number

When a light ray propagates through a volume of air, scattering at the particles of the air (and "foreign" particles floating in it) deflects part of the ray's intensity in all directions and changes its color. As described in Chapter 8 this phenomenon is called extinction. Atmospheric extinction is perhaps best known from the way the setting sun gets dimmer and redder when approaching the horizon, where its light traverses increasingly denser layers of air and encounters more particles that may scatter it. At twilight, the sun's apparent color is also affected by absorption at aerosol particles in the atmosphere.

Provided the scattering particles in a volume of air are much smaller than the wavelength of incident light, Rayleigh's theory can be used to infer the scatterers' number density from the attenuation of sunlight. One may even obtain an estimate of Avogadro's number. In the following we shall ignore the effects of absorption and assume that the scatterers are homogenously distributed throughout the air.

As suggested by Figure 8.4 (page 210), we can expect that the brighter a light ray has been initially, the more it will lose of its intensity by scattering in all directions. In addition, its intensity will decrease in proportion to the distance it crosses through the medium. Let us assume that the decrement in the intensity of a light ray, dI, will be proportional to its initial intensity, I, and the length dx of its path through the air:

$$dI = -hIdx,$$

where the minus sign indicates that the intensity of the light ray is diminished and the factor h, called turbidity, tells us how efficiently the air diverts light.

Dividing both sides by I yields:

$$\frac{dI}{I} = -hdx.$$

Entering the medium with intensity I_0, the ray will diminish to an intensity I after traversing a distance x in the medium. Assuming that h is constant throughout the volume of air under consideration we can integrate

$$\int_{I_0}^{I} \frac{dI}{I} = -\int_{0}^{x} hdx$$

and obtain

$$I(x) = I_0 e^{-hx}$$

Although commonly called Lambert's law (after Johann Heinrich Lambert), this exponential law was first stated by Pierre Bouguer in his 1729 *Essay on the Gradation of Light*, which also describes his measurements of extinction as described on pages 211–12.

In his 1899 paper on light scattering, Lord Rayleigh derived a formula for the turbidity h in the framework of his theory of scattering at small spherical particles, assuming that multiple scattering is negligible:

$$h = \frac{32\pi^3(n-1)^2}{3N\lambda^4}$$

where n is the index of refraction of air (about 1.0003), N is the number of scatterers per unit volume, and λ the wavelength of light.

Thus, measuring the attenuation of light (of known wavelength λ) on its path through the air (with known index of refraction) gives an indication as to the number density of the scatterers. If one accepts the molecular hypothesis, Avogadro's number ensues, as described on pages 214–17 in Chapter 8.

Notes

All translations in the text are by the Götz Hoeppe and John Stewart unless otherwise noted.

PROLOGUE

1. Chuang Tzu, *The Inner Chapters*, trans. Angus Graham (London: Allen & Unwin, 1981), 43.

2. Dietrich Wildung, "Ägyptisch Blau," in *Blau: Farbe der Ferne*, ed. Hans Gercke (Heidelberg:Verlag Das Wunderhorn, 1990), 53.

3. L. Levy-Erell, "Die Ewe: Ein Negerstamm der Goldküste," *Atlantis* (Berlin), no. 6 (1930): 348.

4. Diana Birch, *Ruskin on Turner* (London: Cassell, 1990), 29.

5. Ludwig Friedrich Kämtz, *Lehrbuch der Meteorologie* (Halle: Gebauer'sche Buchhandlung, 1836), 34–35.

6. Leonhard Euler, *Briefe über verschiedene Gegenstände der Naturlehre* (Leipzig: Verlag der Dyck'schen Buchhandlung, 1792), 178–79.

7. Albert Heim, *Luftfarben* (Zurich: Hofer, 1912), 24.

8. Hermann Ethé, *Zakarija ben Muhammed ben Mahmud el-Kazwini's Kosmographie: Die Wunder der Schöpfung* (Leipzig: Fue's Verlag, 1868), 347.

9. Gottfried Wilhelm Muncke, "Atmosphäre," in *Johann Samuel Traugott Gehler's Physikalisches Wörterbuch, neu berarbeitet von Brandes, Gmelin, Horner, Muncke, Pfaff* (Leipzig: E. B. Schwickert, 1825), 1:506.

CHAPTER 1   Of Philosophers and the Color Blue

1. Aratos, *Phaenomena*, trans. Douglas Kidd (Cambridge: Cambridge University Press, 1997).

2. *Odyssey*, bk. 3, lines 1–3; Homer, *The Odyssey*, trans. A. T. Murray (Cambridge, Mass.: Harvard University Press, 1919), 81.

3. *Odyssey*, bk. 5, lines 130–32; Homer, *The Odyssey*, 191–93.

4. W. E. Gladstone, *Homer's Perception and Use of Colour*, vol. 3 in *Studies on Homer and the Homeric Age* (Oxford: Oxford University Press, 1858). See also Elizabeth Henry Bellmer, "The Statesman and the Ophthalmologist: Gladstone and Magnus on the Evolution of Human Colour Vision, One Small Episode of the Nineteenth-Century Darwinian Debate," *Annals of Science* 56 (1999): 25–45.

5. *Iliad*, bk. 24, lines 93–95; Homer, *The Iliad*, trans. A. T. Murray (Cambridge, Mass.: Harvard University Press, 1925), 2:569.

6. *De anima* 418b7–9; Aristotle, *On the Soul, Parva Naturalia, On Breath*, trans. W. S. Hett, Loeb Classical Library (London: Heinemann, 1964), 105.

7. *De sensu* 439b19–28; Aristotle, *On the Soul, Parva Naturalia, On Breath*, 233.

8. *De sensu* 440a7–12; Aristotle, *On the Soul, Parva Naturalia, On Breath*, 233–34.

9. *Cratylus* 396c; *The Dialogues of Plato*, trans. B. Jewett, 4th ed. (Oxford: Clarendon Press, 1953), 3:56. *Phaedrus* 270a; *The Dialogues of Plato*, 3: 178.

10. Aristophanes, *The Acharnians, The Clouds, Lysistrata*, trans. Alan H. Sommerstein (London: Penguin, 1973), 121–22.

11. *Meteorologica* 341b6ff; Aristotle, *Meteorologica*, trans. H. D. P. Lee (Cambridge, Mass.: Harvard University Press, 1952), 21.

12. *Meteorologica* 366b14–19; Aristotle, *Meteorologica*, 209.

13. Hermann Diels and Walther Kranz, *Die Fragmente der Vorsokratiker*, 18th ed. (Zurich/Hildesheim: Weidmann, 1989), 1:95.

14. *Iliad*, bk. 14, line 288; Homer, *The Iliad*, trans. A. T. Murray (Cambridge, Mass.: Harvard University Press, 1925), 2:89.

15. Diels and Kranz, *Die Fragmente der Vorsokratiker*, 2:151; Hermann Schmitz, "Die Luft und als Was Wir Sie spüren," in *Luft*, ed. Bernd Busch (Cologne: Wienand, 2003), 76–84.

16. *Meteorologica* 342b2–11; Aristotle, *Meteorologica*, 37.

17. Theophrastus (Pseudo-Aristotle), *De coloribus* 793b34–794a15; Aristotle, *Minor Works*, trans. W. S. Hett, Loeb Classical Library (Cambridge, Mass.: Harvard University Press, 1952), 19.

CHAPTER 2   A Blue Mixture: Light and Darkness

1. Otto Spies, "Al-Kindi's Treatise on the Cause of the Blue Colour of the Sky," *Journal of the Bombay Branch of the Royal Asiatic Society*, n.s., 13 (1937): 17.

2. *Optics*, bk. 1, chap. 3, sec. 44; A. I. Sabra, *The Optics of Ibn Al-Haytham: Books I–III, On Direct Vision* (London: Warburg Institute, 1989), 1:29.

3. *Optics*, bk. 3, chap. 7, sec. 19; Sabra, *The Optics of Ibn Al-Haytham*, 1:287.

4. Hermann Ethé, *Zakarija ben Muhammed ben Mahmud el-Kazwini's Kosmographie: Die Wunder der Schöpfung*, 1st half-vol. (Leipzig: Fue's Verlag, 1868), 323.

5. Abu al-Rayhan Muhammad Ibn Ahmad al-Biruni, *The Exhaustive Treatise on Shadows*, trans. and commentary by E. S. Kennedy (Aleppo: Institute for the History of Arabic Science, University of Aleppo, 1976), 31.

6. Eilhard Wiedemann, "Ansichten von muslimischen Gelehrten über die blaue Farbe des Himmels," in *Arbeiten aus den Gebieten der Physik, Mathematik, Chemie; Julius Elster und Hans Geitel gewidmet* (Braunschweig: Friedrich Vieweg, 1915), 124–25.

7. Exodus 24, 9–10; *The Bible*, Authorized Version for the Bible Society based on the King James Version (Oxford: Oxford University Press, 1994), 74.

8. Ezekiel 1:26; *The Bible*, 74.

9. Genesis 1:6–7; *The Bible*, 5.

10. *Opus maius*, 9th div., chap. I; Roger Bacon, *The Opus Maius of Roger Bacon*, trans. Robert Belle Burke (Philadelphia: University of Pennsylvania Press, 1928), 2:482.

11. Ibid., 2:482.

12. *On the Composition of the World*, bk. 7, chap. 16; Ristoro d'Arezzo, *Della Composizione del Mondo di Ristoro d'Arezzo* (Milan: G. Daelli e Comp., 1864), 282.

13. Ibid., 283.

14. A comprehensive account of ancient observations of star colors is given by Franz Boll, "Antike Beobachtungen farbiger Sterne," *Abhandlungen der Königlich Bayerischen Akademie der Wissenschaften, Philosophisch-philologische und historische Klasse* (1916), 1. Abhandlung.

CHAPTER 3   Aerial Perspective

1. Carol Vogel, "Leonardo Notebook Sells for $30.8 Million," *New York Times*, November 12, 1994, p. 1; Harry Bellet, "La Longue lignée des propriétaires du Codex Leicester," *Le Monde (Paris)*, February 7, 1997, p. 23. I thank Mr. Stephen C. Massey, former director of books and manuscripts at Christie's, New York, for additional information concerning the auction.

2. *Codex Atlanticus*, fol. 119v; David C. Lindberg, *Theories of Vision from Alkindi to Kepler* (Chicago: University of Chicago Press, 1976), 155.

3. Translated by Samuel Edgerton from a Latin copy of *Della pittura*. Samuel Y. Edgerton, Jr., "Alberti's Color Theory: A Medieval Bottle Without Renaissance Wine," *Journal of the Warburg and Courtauld Institutes* 32 (1969): 122.

4. *Codex Urbinas*, fol. 75v; Martin Kemp, *The Science of Art: Optical Themes in Western Art from Brunelleschi to Seurat* (New Haven, Conn.: Yale University Press, 1990), 268.

5. *Codex Trivulzianus*, fol. 39r; Janis C. Bell, "Color Perspective, c. 1492," *Achademia Leonardi Vinci* 5 (1992): 72.

6. *Bibliotheque Nationale, Cod. BN 2038*, fol. 25v; Janis C. Bell, "Color Perspective, c. 1492," 71.

7. *Bibliotheque Nationale, Cod BN 2038*, fol. 18v; Jean Paul Richter, *The Literary Works of Leonardo da Vinci*, 3rd ed. (London: Phaidon Press, 1883), 1:236–37.

8. MS C, fol. 18r; Edward MacCurdy, *The Notebooks of Leonardo da Vinci*, 2 vols. (London: Jonathan Cape, 1938), 2:296.

9. MS H, fol. 77 [29]v; MacCurdy, *The Notebooks of Leonardo da Vinci*, 1:411.

10. *Codex Leicester*, fol. 4r; MacCurdy, *The Notebooks of Leonardo da Vinci*, 1:418–20.

11. *Codex Leicester*, fol. 36r; MacCurdy, *The Notebooks of Leonardo da Vinci*, 1:420–21.

CHAPTER 4   A Color of the First Order

1. Richard S. Westfall, *Never at Rest: A Biography of Isaac Newton* (Cambridge: Cambridge University Press, 1980), 59.

2. Westfall, *Never at Rest*, 64.

3. Alan E. Shapiro, *The Optical Papers of Isaac Newton*, vol. 1, *The Optical Lectures 1670–1672* (Cambridge: Cambridge University Press, 1984), 10.

4. *Opticks*, bk. 1, pt. 1, prop. 2; Isaac Newton, *Opticks, or A Treatise of the Reflections, Refractions, Inflections & Colours of Light*, based on the 4th ed. (1730) (New York: Dover, 1952), 32–33.

5. *Opticks*, bk. 1, pt. 2, prop. 2, def.; Newton, *Opticks*, 124–25.

6. *Opticks*, bk. 2, pt. 1; Newton, *Opticks*, 199–200.

7. *Opticks*, bk. 2, pt. 3, prop. 5; Newton, *Opticks*, 251.

8. *Opticks*, bk. 2, pt. 3, prop. 7; Newton, *Opticks*, 257.

9. Alan Shapiro, "Newton's Experiments on Diffraction and the Delayed

Publication of the *Opticks*," in *Isaac Newton's Natural Philosophy*, ed. Jed Z. Buchwald and I. Bernard Cohen (Cambridge, Mass.: MIT Press, 2001), 47–76.

10. Westfall, *Never at Rest*, 374.

11. Robert Boyle, *The General History of the Air, Designed and Begun by the Honorable Robert Boyle [1692]*, vol. 12 in *The Works of Robert Boyle* (London: Pickering & Chatto, 2000), 12.

12. Simon Shapin, *The Scientific Revolution* (Chicago: University of Chicago Press, 1996), 40.

13. Johann Caspar Funck, *Liber de coloribus coeli* (Ulm: Daniel Bartholomaei, 1716).

14. J. D. Forbes, "The Colours of the Atmosphere Considered with Reference to a Previous Paper 'On the Colour of Steam under Certain Circumstances.'" *Transactions of the Royal Society of Edinburgh* 14 (1840): 378.

15. Jean-Henri Hassenfratz, "Sur les altérations que la lumière du soleil éprouve en traversant l'atmosphère," *Annales de Chimie* (Paris) 66 (1808): 54–62.

16. Johann Heinrich Lambert, "Sur la perspective aérienne," in *Nouveaux Mémoires de l'Académie Royale des Sciences et des Belles-Lettres* (1774), 75–76.

17. Jean-Antoine Nollet, *Leçons de Physique Expérimentale* (Paris: Hippolyte-Louis Guerin & Louis-François Delatour, 1764), 6:17–19. In wondering about the appearance of Earth as seen from the moon, Nollet's most famous predecessor is the astronomer Johannes Kepler, who considered this theme in his book *Somnium* (1609); see Stephen J. Dick, *Plurality of Worlds: The Origins of the Extraterrestrial Life Debate from Democritus to Kant* (Cambridge: Cambridge University Press, 1982).

18. J. C. P. Erxleben, *Anfangsgründe der Naturlehre*, 4th ed. (Göttingen: Johann Christian Dieterich, 1787), 306.

19. Leonhard Euler, *Briefe über verschiedene Gegenstände aus der Naturlehre* (Leipzig: Verlag der Dyckschen Buchhandlung, 1792), 1:180.

20. Humphry Davy, "An Essay on Heat, Light and the Combinations of Light," in *The Collected Works of Sir Humphry Davy*, vol. 2, *Early Miscellaneous Papers*, ed. John Davy (London: Smith, Elder and Co, 1839), 3–86.

21. Davy, "An Essay on Heat," 29–30.

CHAPTER 5   Basic Phenomenon, or Optical Illusion?

1. Alexander von Humboldt, "Ueber einen Versuch den Gipfel des Chimborazo zu ersteigen," in *Kleinere Schriften* (Stuttgart: J. G. Cotta'scher Verlag, 1853), 1:150.

2. The cyanometer readings taken during the Atlantic crossing are published in Alexander von Humboldt, "Couleur azurée du ciel et couleur de la mer a sa surface," in *Relations Historiques aux Régions Équatoriales du Nouveau Continent* (Paris: F. Schoell, 1814), 1:248–56. A summary of the cyanometer recordings is available in Humboldt's *Ideen zu einer Geographie der Pflanzen* (Leipzig: Akademische Verlagsgesellschaft, 1960 [1807]), 110–12.

3. Humboldt, "Ueber einen Versuch," 156.

4. Humboldt, *Ideen zu einer Geographie der Pflanzen*, 176.

5. Georg Wilhelm Muncke, "Ueber subjective Farben und gefärbte Schatten," *Journal für Chemie und Physik* 30 (1820): 81.

6. Johann Wolfgang von Goethe, *Zur Naturwissenschaft überhaupt, besonders zur Morphologie, Erfahrung, Betrachtung, Folgerung, durch Lebensereignisse verbunden*, vol. 12 in *Sämtliche Werke nach Epochen seines Schaffens, Münchner Ausgabe* (Munich: Hanser-Verlag, 1989), 570.

7. Idem, 570.

8. *Maximen und Reflexionen* (1828), §575; Johann Wolfgang von Goethe, *Wilhelm Meisters Wanderjahre, Maximen und Reflexionen*, vol. 17 in *Sämtliche Werke nach Epochen seines Schaffens, Münchner Ausgabe* (Munich: Hanser Verlag, 1991), 824.

9. Johann Wolfgang von Goethe, *Italienische Reise*, vol. 15 in *Sämtliche Werke nach Epochen seines Schaffens, Münchner Ausgabe* (Munich: Hanser Verlag, 1992), 102.

10. Idem, 609.

11. Description in Goethe's *Tag- und Jahres-Heften für 1790*; Johann Wolfgang von Goethe, "Zur Farbenlehre," vol. 10 in *Sämtliche Werke nach Epochen seines Schaffens, Münchner Ausgabe* (Munich: Hanser Verlag, 1989), 1004.

12. Letter to Carl Friedrich von Zelter dated June 22, 1808; Goethe, "Zur Farbenlehre," 998.

13. *Enthüllung der Theorie Newtons: Des ersten Bandes zweiter, polemischer Teil*; Goethe, "Zur Farbenlehre," 275–472.

14. Marjorie Hope Nicholson, *Newton Demands the Muse: Newton's Opticks and Eighteenth Century Poets* (Princeton, N.J.: Princeton University Press, 1946).

15. *Zur Farbenlehre*, Didaktischer Teil, § 154–55; Goethe, "Zur Farbenlehre," 69.

16. Werner Heisenberg, "Das Naturbild Goethes und die technisch-wissenschaftliche Welt," *Jahrbuch der Goethe-Gesellschaft* (Frankfurt am Main), n.s., 29 (1967): 27–42.

17. Gottfried Wilhelm Muncke, "Atmosphäre," in *Johann Samuel Traugott Gehlers Physikalisches Wörterbuch, neu bearbeitet von Brandes, Gmelin, Horner, Muncke, Pfaff* (Leipzig: E. B. Schwickert, 1825), 1:454.

CHAPTER 6   A Polarized Sky

1. Ernst Wilhelm von Brücke, "Ueber die Farben, welche trübe Medien im auffallenden und durchfallenden Lichte zeigen," *Annalen der Physik und Chemie*, 3rd ser., 28 (1853): 382.

2. Idem, 384.

3. Jacques Babinet, "Sur un nouveau point neutre dans l'atmosphère," *Comptes Rendus* (Paris) 11 (1840): 618–20.

4. Quoted in John Tyndall, "On the Blue Colour of the Sky, the Polarization of Skylight, and on the Polarization of Light by Cloudy Matter Generally," *Philosophical Magazine*, 4th ser., 37 (1869): 389.

5. Idem, 385.

6. John Tyndall, "On the Action of Rays of High Refrangibility upon Gaseous Matter." *Philosophical Transactions of the Royal Society of London* 160 (1870): 347.

7. John Tyndall, "On a New Series of Chemical Reactions Produced by Light," *Proceedings of the Royal Society of London* 17 (1868): 97.

8. Tyndall, "On the Action of Rays," 347.

9. Rudolf Clausius, "Ueber das Vorhandenseyn von Dampfbläschen in der Atmosphäre und ihren Einfluss auf die Lichtreflexion und die Farben derselben," *Annalen der Physik und Chemie*, 3rd ser., 28 (1853): 556.

CHAPTER 7   Lord Rayleigh's Scattering

1. The draft of Maxwell's lecture is reprinted in P. M. Harman, ed., *The Scientific Letters and Papers of James Clerk Maxwell, Vol. I: 1846–1862* (Cambridge: Cambridge University Press, 1990), 675–79.

2. John William Strutt, "Some Experiments on Colour," *Nature* 3 (1871): 235.

3. John William Strutt, "On the Light from the Sky, Its Polarization and Colour," *Philosophical Magazine*, 4th ser., 41 (1871): 107.

4. Idem, 111.

5. The spectral distribution of skylight had been measured three years previously by A. de la Rive, "Note sur un photomètre destiné a mesurer la transparence de l'air," *Annales de Chimie et de Physique* (Paris), 4th ser., 12 (1867): 243–49.

6. Quoted by Robert John Strutt, *Life of John William Strutt, Third Baron Rayleigh*, 2nd ed. (Madison: University of Wisconsin Press, 1968 [1924]), 54.

7. A. Lallemand, "Sur la polarisation et la fluorescence de l'atmosphère," *Comptes Rendus* (Paris) 75 (1872): 707–11.

8. George Gabriel Stokes, "On Change of Refrangibility of Light," *Philosophical Transactions of the Royal Society of London* (1852): pt. 1, 463–562.

9. Quoted by Francis Everitt, "James Clerk Maxwell," in *Dictionary of Scientific Biography* (New York: Charles Scribner's Sons, 1974), 9:210.

10. This comparison is made by Craig F. Bohren and Donald R. Huffman, *Absorption and Scattering of Light by Small Particles* (New York: John Wiley, 1983), 10.

CHAPTER 8   Molecular Reality

1. Rayleigh quotes Maxwell's letter in his paper, "On the Transmission of Light Through an Atmosphere Containing Small Particles in Suspension," *Philosophical Magazine*, 5th ser., 47 (1899): 376. The letter is reprinted at full length in P. M. Harman, *The Scientific Letters and Papers of James Clerk Maxwell, Vol. II: 1862–1873* (Cambridge: Cambridge University Press, 1995): 919–20.

2. Pierre Bouguer, *Traité d'Optique sur la Gradation de la Lumière* (Paris: Imprimerie de H. L. Guerin & L. F. Delatour, 1760).

3. John William Strutt (third Lord Rayleigh), "On the Transmission of Light," 382.

4. Ludwig Valentin Lorenz, "Lysbevaegelsen i og uden for en af plane Lysbolger belyst Kugle," *Det Kongelige Danske Videnskabernes Selskabs Skrifter*, no. 6, *Naturvidenskabelig og mathematisk Afdeling* 6 (1890): 1–62.

5. Albert Einstein, "Über die von der molekularkinetischen Theorie der Wärme geforderte Bewegung von in ruhenden Flüssigkeiten suspendierten Teilchen," *Annalen der Physik* (Leipzig), 4th ser., 17 (1905): 549–60; idem., "Zur Theorie der Brownschen Bewegung," *Annalen der Physik* (Leipzig), 4th ser., 19 (1905): 371–81.

6. Quoted by Mary Jo Nye, *Molecular Reality: A Perspective on the Scientific Work of Jean Perrin* (London/New York: Macdonald/American Elsevier, 1972), 161.

7. Albert Einstein, "Theorie der Opaleszenz von homogenen Flüssigkeiten und Flüssigkeitsgemischen in der Nähe ihres kritischen Zustands," *Annalen der Physik* (Leipzig), 4th ser., 33 (1910): 1295.

8. Rayleigh reiterates this viewpoint much later in "Colours of Sea and Sky," *Nature* 83 (1910): 48–50.

9. C. V. Raman, "The molecular scattering of light," in *Nobel Lectures Physics 1922–1941*, 267–75 (Amsterdam: Elsevier, 1965), 267–75.

10. P.J.E. Peebles and J. T. Yu, "Primeval Adiabatic Perturbations in an Expanding Universe," *Astrophysical Journal* 162 (1970): 815–36; Qingjuan Yu, David N. Spergel, and Jeremiah P. Ostriker, "Rayleigh Scattering and Microwave Background Fluctuations," *Astrophysical Journal* 558 (2001): 23–28.

11. Cited by R. B. Lindsay, "John William Strutt, third Baron Rayleigh," in *Dictionary of Scientific Bibliography* (New York: Charles Scribner's Sons), vol. 13 (1976), 105.

CHAPTER 9   Ozone's Blue Hour

1. Angelika Lochmann and Angelika Overath, *Das blaue Buch: Lesarten einer Farbe* (Nördlingen: Franz Greno, 1982), 208–9.

2. William Least Heat Moon, *Blue Highways: A Journey into America* (Boston: Little, Brown, 1982).

3. A. Houzeau, "Preuve de la présence dans l'atmosphère d'un nouveau principe gazeux, l'oxygène naissant," *Comptes Rendus* (Paris), 46 (1858): 89–91.

4. A. Cornu, "Sur la limite ultra-violette du spectre solaire," *Comptes Rendus* (Paris) 88 (1879): 1101–8; ibid., "Sur l'absorption par l'atmosphère des radiations ultra-violettes," *Comptes Rendus* (Paris) 91 (1879): 1285–90.

5. P. Hautefeuille and J. Chappuis, "Sur la liquéfaction de l'ozone et sur sa couleur à l'état gazeux," *Comptes Rendus* (Paris) 91 (1880): 522–25.

6. C. Fabry and H. Buisson, "Ètude de l'extrémitè ultra-violette du spectre solaire," *Journal de Physique et le Radium*, ser. 6, vol. 2 (1921): 197–226.

7. S. Chapman, "On Ozone and Atomic Oxygen in the Upper Atmosphere," *Philosophical Magazine*, 7th ser., 10 (1930): 369–83.

8. E. O. Hulburt, "The Brightness of the Twilight Sky and the Density and Temperature of the Atmosphere," *Journal of the Optical Society of America* 28 (1938): 227–36.

9. E. O. Hulburt, "The Upper Atmosphere of the Earth," *Journal of the Optical Society of America* 37 (1947): 412.

10. Heiner Flöthmann, Christian Beck, Reinhard Schinke, Clemens Woywod, and Wolfgang Domcke, "Photo-dissociation of ozone in the Chappuis band. II. Time-dependent wave-packet calculations and interpretation of diffuse vibrational structures," *Journal of Chemical Physics* 107 (1997): 7296–313.

11. Edwin Olson Hulburt, "Some Recent Papers in the *Journal of the Optical Society of America*," *Journal of the Optical Society of America* 46 (1956): 9.

12. Aden Meinel and Marjorie Meinel, *Sunsets, Twilights and Evening Skies* (Cambridge: Cambridge University Press, 1983), chap. 4; David K. Lynch and William Livingston, *Color and Light in Nature* (Cambridge: Cambridge University Press, 2001), 33–45.

13. Jean Dubois, "L'ombre de la Terre," *Comptes Rendus* (Paris) 222 (1946): 671–72; ibid., "Résultats de nouvelles recherches sur l'Ombre de la Terre," *Comptes Rendus* (Paris) 226 (1948): 1180–83.

14. A computer-based demonstration of the Purkinje phenomenon can be found in Peter Kaiser's Web book, *The Joy of Visual Perception* (www.yorku. ca/eye/).

15. J. C. Farman, B. G. Gardiner, and J. D. Shanklin, "Large losses of total ozone in Antarctica reveal seasonal ClO_x/NO_x interaction," *Nature* 315 (1985): 207–10.

16. This situation is now changing. Upon my request, Wolfgang Meyer, in 2005 head of Neumayer station, regularly took color photographs of the twilight sky throughout his residence in Antarctica. Professor Raymond Lee of the United States Naval Academy is about to analyze these images for the anticipated color change during maximal ozone depletion.

CHAPTER 10   The Color of Life

1. Jean Baptiste Joseph Fourier, "Remarques Générales sur les Températures du globe terrestre et des espaces planétaires," *Annales de Chimie et de Physique* (Paris), 27 (1824) 136–67.

2. John Tyndall, "On the absorption and radiation of heat by gases and vapours, and on the physical connexion of radiation, absorption and conduction," *Philosophical Magazine*, 4th ser., 22 (1861): 276–77.

3. Quoted by Vaclav Smil, *The Earth's Biosphere: Evolution, Dynamics and Change* (Cambridge, Mass.: MIT Press, 2002), 4.

4. V. I. Vernadsky, *The Biosphere* (New York: Copernicus Press, 1998), 44.

5. Cited by Gale E. Christensen, *Greenhouse: The 200-year Story of Global Warming* (New York: Walker and Co., 1999), 115.

6. G. S. Callendar, "The Artificial Production of Carbon Dioxide and Its Influence on Temperature," *Quarterly Journal of the Royal Meteorological Society* 64 (1938): 223–37.

7. Roger Revelle and Hans S. Suess, "Carbon Dioxide Exchanges Between Atmosphere and Ocean and the Question of an Increase of Atmospheric CO_2 During the Past Decades," *Tellus* 9 (1957): 19–20.

8. J. R. Petit, J. Jouzel, D. Raynaud, et al., "Climate and Atmospheric History of the past 420,000 Years from the Vostok Ice Core, Antarctica," *Nature* 399 (1999): 429–36.

9. For a summary of the evidence for climatic change in the twentieth century, see John Houghton, *Global Warming: The Complete Briefing*, 3rd ed. (Cambridge: Cambridge University Press, 2004), chap. 4.

10. James F. Kasting and David Catling, "Evolution of a Habitable Planet," *Annual Review of Astronomy and Astrophysics* 41 (2003): 443.

EPILOGUE

1. Dezső Tandori, *Birds and Other Relations: Selected Poetry of Dezső Tandori*, trans. Bruce Berlind (Princeton, N.J.: Princeton University Press, 1986), 21.

2. Albert Einstein, "Ernst Mach." *Physikalische Zeitschrift* 17 (1916): 102.

Further Reading

General References

Gage, John. *Color and Culture: Practice and Meaning from Antiquity to Abstraction*. Boston: Little, Brown and Co., 1993. (A standard work by an art historian, containing many references.)

Gercke, Hans, ed. *Blau: Farbe der Ferne*. Heidelberg:Verlag Das Wunderhorn, 1990. (A comprehensive volume on the color blue in art and history; in German.)

Greenler, Robert. *Rainbows, Halos and Glories*. Cambridge: Cambridge University Press, 1980. (A field guide for the naked-eye exploration of the atmosphere.)

Kemp, Martin. *The Science of Art: Optical Themes in Western Art from Brunelleschi to Seurat*. New Haven, Conn.: Yale University Press, 1990. (A fascinating overview of the art and science of perspective and color.)

Lynch, David K., and William Livingston. *Color and Light in Nature*. 2nd ed. Cambridge: Cambridge University Press, 2001. (An updated sequel to Minnaert's book, including concise explanations and many stunning photographs.)

Minnaert, Marcel. *The Nature of Light and Colour in the Open Air*. New York: Dover Publications, 1954. (The classic work on visual observation of the daytime sky, it has inspired and accompanied generations of observers.)

Pastoureau, Michel. *Blue: The History of a Color*. Princeton, N.J.: Princeton University Press, 2001. (A cultural history of the color blue, with many references.)

CHAPTER 1   Of Philosophers and the Color Blue

Aristotle. *Meteorology*. Trans. H.D.P. Lee. Cambridge, Mass.: Harvard University Press, 1952. (One of several translations; available in larger academic libraries.)

Barnes, J., ed. *The Cambridge Companion to Aristotle*. Cambridge: Cambridge University Press, 1995.

Boyer, Carl B. *The Rainbow: From Myth to Mathematics*. Princeton, N.J.: Princeton University Press, 1987 [1959]. (Comprehensive study of historical explanations of the rainbow.)

Gladstone, W. E. *Homer's perception and use of colour*. Vol. 3 of *Studies on Homer and the Homeric Age*. Oxford: Oxford University Press, 1858.

Gottschalk, H. B. "The De Coloribus and Its Author." *Hermes* 92 (1964): 59–85. (Argues for Theophrastus as the author of *De coloribus*.)

Guerlac, Henri. "Can there be colors in the dark? Physical color theory before Newton." *Journal of the History of Ideas* 47 (1986): 3–20. (A fresh critical review of a classical problem.)

Lee, Raymond L., Jr., and Alistair Fraser. *The Rainbow Bridge: Rainbows in Art, Myth, and Science*. University Park: University of Pennsylvania Press, 2001. (A comprehensive monograph on the rainbow, with up-to-date treatment of the relevant physics and colorimetry.)

Lindberg, David C. *The Beginnings of Western Science: The European Tradition in Philosophical, Religious and Institutional Context, 600 B.C. to A.D. 1450*. Chicago: University of Chicago Press, 1992. (Contains an overview of Greek cosmology and puts Greek science into its social context.)

Lyons, John. "Color in Language." In *Colour: Art & Science*, edited by Trevor Lamb and Jeanne Bourriau, 194–224. Cambridge: Cambridge University Press, 1995. (A succinct summary of the debate on the meaning of color terms.)

Taub, Liba. *Ancient Meteorology*. London: Routledge, 2003. (A very good overview of the subject, focusing on rules for weather prediction.)

CHAPTER 2   A Blue Mixture: Light and Darkness

Bacon, Roger. *The "Opus Majus" of Roger Bacon*. Translated from the Latin by Robert Belle Burke. 2 vols. Philadelphia: University of Pennsylvania Press, 1928.

Lindberg, David C. *Theories of Vision from Al-Kindi to Kepler*. Chicago: University of Chicago Press, 1976. (Comprehensive source for the optical tradition of the Islamic and Christian Middle Ages.)

Meier, Christel. *Gemma Spiritalis: Methode und Gebrauch der Edelsteinallegorese vom frühen Christentum bis ins 18. Jahrhundert*. Munich: Wilhelm Fink Verlag,

1977. (A detailed account of the allegoresis of gems, focusing on the Middle Ages.)

Ribémont, Bernard, ed. *Observer, Lire, Écrire le Ciel au Moyen Age*. Paris: Klinck-sieck, 1991.

Ristoro d'Arezzo. *Della Composizione del Mondo*. Milan: G. Daelli e Comp., 1864 [1282].

Sabra, A. I. *The Optics of Ibn Al-Haytham: Books I–III, On Direct Vision*. 2 vols. London: The Warburg Institute, 1989. (Translation of al-Haytham's major work.)

Sauvanon, Jeanine. *La Cathédrale de Chartres: Miroir de la Nature*. Le Coudray: Éditions Legué-Houvet, 2004. (Illustrated guide to the stained glass windows of Chartres Cathedral.)

Spies, Otto. "Al-Kindi's treatise on the cause of the blue colour of the sky." *Journal of the Bombay Branch of the Royal Asiatic Society*, n.s., 13 (1937): 7–19. (A revised translation, relying mostly on the Oxford manuscript of al-Kindi's treatise.)

Turner, H. R. *Science in Medieval Islam*. Austin: University of Texas Press, 1995. (Illustrated introduction to Arab science, good for a first overview.)

Wiedemann, Eilhard. "Anschauungen von muslimischen Gelehrten über die blaue Farbe des Himmels." In *Arbeiten aus den Gebieten der Physik, Mathematik, Chemie, Julius Elster und Hans Geitel gewidmet*, 118–126. Brunswick: Friedrich Vieweg Verlag, 1915. (First paper relating the rediscovery of the Istanbul manuscript of al-Kindi's treatise on the blue sky.)

CHAPTER 3   Aerial Perspective

Arasse, Daniel. *Leonard de Vinci: Le rhytme du monde*. Paris: Hazan, 1997. (Monumental study with a critical assessment of Leonardo's scholarly development; after much praise for Leonardo as an other-worldly genius, Arasse succeeds in demonstrating how much he was a child of his times.)

Bell, Janis. "Color Perspective, c. 1492." *Achademia Leonardi Vinci* 5 (1992): 64–77.

Bell, Janis. "Aristotle as a Source for Leonardo's Theory of Colour Perspective after 1500." *Journal of the Warburg and Courtauld Institutes* 56 (1993): 100–18. (Two fascinating papers that reveal how Leonardo increasingly turned to Aristotelian concepts of optics and cosmology.)

Edgerton, Samuel Y., Jr. "Alberti's Colour Theory: A Medieval Bottle without Renaissance Wine." *Journal of the Warburg and Courtauld Institutes* 32 (1969): 109–34. (A critical account of Alberti's color theory.)

Hall, Marcia B. *Color and Meaning: Practice and Theory in Renaissance Painting.* Cambridge: Cambridge University Press, 1992. (Comprehensive account of Renaissance theories of color and their impact on the practices of painters from Duccio to Tintoretto.)

Harrison, Edward R. *Darkness at Night: A Riddle of Cosmology.* Cambridge, Mass.: Harvard University Press, 1987. (A fascinating and accessible account.)

Kemp, Martin. *Leonardo da Vinci: The Marvellous Works of Nature and Man.* London: J. M. Dent & Sons, 1981. (Useful introduction to Leonardo's varied works.)

MacCurdy, Edward, ed. *The Notebooks of Leonardo da Vinci.* 2 vols. London: Jonathan Cape, 1938.

Richter, Jean Paul, ed. *The Literary Works of Leonardo da Vinci.* 3rd ed. London: Phaidon Press, 1970 [1883]. (Accessible compendium of Leonardo's writings; criticized by recent scholarship for misrepresenting Leonardo's train of thought by splitting it into many subject matters.)

CHAPTER 4 A Color of the First Order

Boyle, Robert. *The General History of the Air.* Vol. 12 in *The Works of Robert Boyle.* London: Pickering & Chatto, 2000 [1692].

Claudius, Rudolf. "Ueber das Vorhandenseyn von Dampfbläschen in der Atmosphäre und ihren Einfluß auf die Lichtreflexion und die Farben derselben." *Annalen der Physik und Chemie*, 3rd ser., 28 (1853): 543–56. (A mid-nineteenth century Newtonian explanation of the blue color of the sky, invoking bubbles of water floating in the air; this hypothesis was soon ruled out by observations.)

Forbes, J. D. "The Colours of the Atmosphere Considered with Reference to a Previous Paper 'On the Colour of Steam under Certain Circumstances.'" *Transactions of the Royal Society of Edinburgh* 14 (1840): 375–91. (A nineteenth-century review containing a critical evaluation of Newton's theory.)

Newton, Isaac. *Opticks, or A Treatise of the Reflections, Refractions, Inflections & Colours of Light.* Reprint of the 4th ed. (1730). New York: Dover Publications, 1952. (One of several reprints of what is probably Newton's most accessible book.)

Sabra, A. I. "The Authorship of the Liber de Crepusculis, an Eleventh-Century Work on Atmospheric Refraction." *Isis* 58 (1967): 77–85 (A medieval treatise on twilight, with an estimate of the height of the atmosphere; until then wrongly ascribed to Ibn al-Haytham.)

Sabra, A. I. *Theories of Light from Descartes to Newton*. London: Oldbourne, 1967. (A detailed survey of optics in the seventeenth century.)

Shapin, Steven. *The Scientific Revolution*. Chicago: University of Chicago Press, 1997.

Shapin, Steven, and Simon Schaffer. *Leviathan and the Air-Pump: Boyle, Hobbes, and the Experimental Life*. Princeton, N.J.: Princeton University Press, 1985. (An account of the seventeenth-century discussion on the mechanical properties of air; already a classic in the burgeoning field of science studies.)

Shapiro, Alan E. *The Optical Papers of Isaac Newton*. Vol. 1, *The Optical Lectures 1670–1672*. Cambridge: Cambridge University Press, 1984.

Shapiro, Alan E. *Fits, Passions, and Paroxysms: Physics, Method, and Chemistry and Newton's Theories of Colored Bodies and Fits of Easy Reflection*. Cambridge: Cambridge University Press, 1993. (Detailed scholarly account of Newton's theory of the colors of bodies and the reactions of his critics up to the mid-19th Century.)

Westfall, Richard S. "Isaac Newton's Coloured Circles twixt two Contiguous Glasses." *Archive for History of Exact Science* 2 (1965): 181–96. (Study of an unpublished manuscript of Newton describing his first measurements of the colored rings.)

Westfall, Richard S. *Never at Rest: A Biography of Isaac Newton*. Cambridge: Cambridge University Press, 1980. (The standard biography of Newton.)

CHAPTER 5   Basic Phenomenon, or Optical Illusion?

Beguelin, N., de. "Sur les Ombres Colorées." *Histoire de l'Académie Royale des Sciences et Belles Lettres (Berlin)*, (1767): 27–40. (An important eighteenth-century paper on colored shadows.)

Brandes, H. W. "Kyanometer." In *Johann Samuel Traugott Gehler's Physikalisches Wörterbuch, neu bearbeitet von Brandes, Gmelin, Horner, Muncke, Pfaff*, vol. 5, 1367–72. Leipzig: E. B. Schwickert, 1829.

Churma, Michael E. "Blue shadows: Physical, physiological and psychological causes." *Applied Optics* 33 (1994): 4719–22. (Overview of the modern interpretation of colored shadows.)

Goethe, John Wolfgang v. *Goethe's Theory of Colors*. Translated by Charles Lock Eastlake. London: John Murray, 1840 [1810]. (There are several English editions of Goethe's book, but Eastlake's translation remains the classic and has been reprinted in abridged versions.)

Humboldt, Alexander von, and Aimé Bonpland. *Ideen zu einer Geographie der Pflanzen nebst einem Naturgemälde der Tropenländer.* Leipzig: Akademische Verlagsgesellschaft, 1960 [1807]. (Humboldt's summary of his measurements with the cyanometer.)

Humboldt, Alexander von. "Ueber einen Versuch den Gipfel des Chimborazo zu ersteigen." Vol. 1 in *Kleinere Schriften*, 133–57. Stuttgart: J. G. Cotta'scher Verlag, 1853. (Humboldt's account of his ascent of Chimborazo.)

Muncke, G. W. "Ueber subjective Farben und gefärbte Schatten." *Journal für Chemie und Physik* 30 (1820): 74–88. (Paper containing Muncke's assertion that the sky's blue color is an optical illusion.)

Reid, Neil, and Friedrich Steinle. "Exploratory Experimentation: Goethe, Land, and Color Theory." *Physics Today* 55 (July 2002): 43–49. (A portrayal of Goethe's experimental technique as a forerunner of the methods later employed in electromagnetism by Faraday and Ampère.)

Saussure, H. B. de. "Description d'un cyanomètre ou d'un appareil destiné à mesurer la transparence de l'air." *Memorie della Accademia delle scienze di Torino* 4 (1788–89): 409–25. (Saussure's original description of the cyanometer.)

Sepper, Dennis L. *Goethe contra Newton: Polemics and the Project for a New Science of Color.* Cambridge: Cambridge University Press, 1988. (A useful introduction that places Goethe's critique of Newton's optics in context.)

CHAPTER 6   A Polarized Sky

Bohren, Craig F. *Clouds in a Glass of Beer: Simple Experiments in Atmospheric Physics.* New York: John Wiley and Sons, 1987. (An entertaining book that includes a good, brief introduction to methods for observing skylight polarization.)

Brücke, Ernst-Wilhelm von. "Ueber die Farben, welche trübe Medien im auffallenden und durchfallenden Lichte zeigen." *Annalen der Physik und Chemie*, 3rd ser., 28 (1853): 363–85. (Brücke's paper on turbid media.)

Brücke, Ernst-Wilhelm von. *Untersuchungen über den Farbenwechsel des africanischen Chamäleons.* Leipzig: Wilhelm Engelmann, 1893. (Brücke's account of his investigation of the chameleon's color changes, contains extensive historical notes.)

Buchwald, Jed Z. *The Rise of the Wave Theory of Light: Optical Theory and Experiment in the Early Nineteenth Century.* Chicago: University of Chicago Press, 1989.

Coulson, K. L. *Polarization and Intensity of Light in the Atmosphere*. Hampton: A. Deepak Publishing, 1988. (Contains a good historical review of early studies of sky polarization and presents modern measurements of skylight intensity and polarization in great detail.)

Hey, J. D. "From Leonardo to the Graser: Light Scattering in Historical Perspective. Part I." *South African Journal of Science* 79 (1983): 1–27. (Useful review of early work on light scattering, focusing on Tyndall.)

Können, G. P. *Polarized Light in Nature*. Cambridge: Cambridge University Press, 1985. (A comprehensive manual for visual observers, includes a section on how to see Haidinger's brush.)

Lipson, S. G., H. Lipson, and D. S. Tannhauser. *Optical Physics*. 3rd ed. Cambridge: Cambridge University Press, 1995. (A readable introduction to optics with emphasis on modern aspects.)

Park, David. *The Fire Within the Eye: A Historical Essay on the Nature and the Meaning of Light*. Princeton, N.J.: Princeton University Press, 1997. (An entertaining history of light; unfortunately, polarization is omitted.)

Roslund, C., and C. Beckman. "Disputing Viking navigation by polarized skylight." *Applied Optics* 33 (1994): 4754–55.

Tyndall, John. "On the Blue Colour of the Sky, the Polarization of Skylight, and on the Polarization of Light by Cloudy Matter generally." *Philosophical Magazine*, 4th ser., 37 (1869): 384–94.

Tyndall, John. "On the Action of Rays of high Refrangibility upon Gaseous Matter." *Philosophical Transactions of the Royal Society of London* 160 (1870): 333–65.

Tyndall, John. *Six Lectures on Light, Delivered in America in 1872–1873*. London: Longmans, Green, and Co., 1873. (Tyndall's popular lectures; vivid descriptions with many metaphors and analogies.)

Wehner, Rüdiger. "Polarized-Light Navigation by Insects." *Scientific American* 235 (July 1976): 106–15. (A review of Wehner's early work on polarized light perception in desert ants.)

CHAPTER 7   Lord Rayleigh's Scattering

Bohren, Craig F. *Clouds in a Glass of Beer: Simple Experiments in Atmospheric Physics*. New York: John Wiley and Sons, 1987. (Describes simple, but often thought-provoking everyday observations and experiments on multiple scattering and polarization.)

Bohren, Craig F. "Atmospheric Optics." In *Encyclopedia of Applied Physics*. Edited by G. L. Trigg. Vol. 12, 405–34. Weinheim: VCH Verlagsgesellschaft, 1995.

Bohren, Craig F., and Alistair Fraser. "Colors of the Sky." *Physics Teacher* 23 (1985): 267–72.

Chandrasekhar, Subramonyam. *Radiative Transfer*. Oxford: Clarendon Press, 1950. (A highly sophisticated treatment by an eminent astrophysicist on light scattering and the transfer of energy in atmospheres, from those of the planets to those of the stars.)

Daston, Lorraine. "The Cold Light of Facts and the Facts of Cold Light: Luminescence and the Transformation of the Scientific Fact, 1600–1750." *EMF—Early Modern France* 3 (1997): 17–44. (Identifies a historical watershed in the interpretation of luminescence—and of scientific "facts" more generally.)

Hartley, Walter N. "On the Limit of the Solar Spectrum, the Blue of the Sky and the Fluorescence of Ozone." *Nature* 39 (1889): 474–77.

Hey, J. D. "From Leonardo to the Graser: Light Scattering in Historical Perspective. Part II." *South African Journal of Science* 79 (1983): 310–24. (Useful sequel to Hey's paper mentioned in the references to Chapter 6.)

Lallemand, A. "Sur la polarisation et la fluorescence de l'atmosphère." *Comptes Rendus Hebdomaires de l'Academie des Sciences de Paris* 75 (1872): 707–11.

Lipson, S. G., H. Lipson, and D. S. Tannhauser. *Optical Physics*. 3rd ed. Cambridge: Cambridge University Press, 1995. (A very good introduction to wave optics, includes modern topics such as critical point opalescence.)

Mahajan, Sanjoy, E. S. Phinney, and Peter Goldreich. *Order of Magnitude Physics: The Art of Approximation in Science* (forthcoming). (Dimensional analysis remains as important a tool today as it was for Rayleigh; this text has many stunning examples demonstrating its power.)

Maxwell, James Clerk. "Experiments on Colour, as perceived by the Eye, with Remarks on Colour-Blindness." *Proceedings of the Royal Society of Edinburgh* 21 (1854): 275–98. (Maxwell's account of his experiments with the color tops and his invention of colorimetry.)

Sherman, Paul. *Colour Vision in the Nineteenth Century: The Young-Helmholtz-Maxwell-Theory*. Bristol: Adam Hilger, 1981. (A summary of the work of nineteenth-century pioneers in the study of color vision.)

Stokes, George Gabriel. "On the Composition and Resolution of Streams of Polarized Light from different Sources." *Transactions of the Cambridge Philosophical Society* 9 (1853): 399–416. (Rayleigh's explanation of skylight polarization is a direct application of the argument made by Stokes in this paper.)

Strutt, John William. "Some Experiments on Colour." *Nature* 3 (1871): 234–36. (Rayleigh's account of his experiments with colored disks.)

Strutt, John William. "On the Light from the Sky, Its Polarization and Colour." *Philosophical Magazine*, 4th ser., 41 (1871a): 107–20, 274–79. (If there is one classic paper on the color of the sky, this is it.)

Strutt, John William. "On the Scattering of Light by small Particles." *Philosophical Magazine*, 4th ser., 41 (1871b): 447–54. (Sequel to the aforementioned paper.)

Strutt, John William (= Lord Rayleigh). "On the Transmission of Light through an Atmosphere containing Small Particles in Suspension." *Philosophical Magazine*, 5th ser., 47 (1899): 375–84. (Contains Rayleigh's derivation of equations of light scattering in a Maxwellian framework, and the hypothesis of air molecules as the atmospheric light scatterers.)

Strutt, Robert John. *Life of John William Strutt, Third Baron Rayleigh*. London: Edward Arnold & Co., 1924. (The fourth baron writing about his father, the third lord.)

Young, A. T. "Rayleigh Scattering." *Physics Today* 35 (January 1982): 42–48. (This paper clarifies different uses of the term scattering, contrasting Rayleigh scattering with the Raman effect.)

Young, Thomas. "On the Theory of Light and Colours." *Philosophical Transactions of the Royal Society of London* 92 (1802): 12–48. (Original description of the tristimulus theory of color vision; first presented at Royal Society's Bakerian lecture of 1801.)

CHAPTER 8   Molecular Reality

Einstein, Albert. "Theorie der Opaleszenz von homogenen Flüssigkeiten und Flüssigkeitsgemischen in der Nähe ihres kritischen Zustands." *Annalen der Physik (Leipzig)*, 4th vol., 33 (1910): 1275–98. (Paper containing Einstein's derivation of the inverse fourth power law from assumptions entirely different from those made by Rayleigh.)

Fowle, Frederick E. "Avogadro's Constant and Atmospheric Transparency." *Astrophysical Journal* 40 (1914): 435–42. (An astronomer's attempt to determine Avogadro's number from atmospheric extinction.)

Loschmidt, Johann Joseph. "Zur Grösse der Luftmoleküle." *Sitzungsberichte der kaiserlichen Akademie der Wissenschaften (Vienna), Mathematisch-naturwissenschaftliche Classe*, 52 (1866): 395–413. (Loschmidt's paper on the size of air molecules.)

Nye, Mary Jo. *Molecular Reality: A Perspective on the Scientific Work of Jean Perrin.* London/New York: Macdonald/American Elsevier, 1972. (Readable account of Perrin's scientific work in the contemporary context.)

Pais, Abraham. *Subtle Is the Lord: The Science and the Life of Albert Einstein.* Oxford: Oxford University Press, 1982. (A scientific biography of Einstein for readers with a strong background in physics; includes a detailed account of his work on critical opalescence.)

Perrin, Jean. "Mouvement Brownien et Réalité Moléculaire." *Annales de Chimie et de Physique,* ser. 8, 18 (1909): 1–109.

Perrin, Jean: *Atoms.* Translated by D. L. Hammick. London: Constable, 1916 [1914].

Renn, Jürgen. "Einstein's Invention of Brownian Motion." *Annalen der Physik* (Leipzig) 14, suppl. (2005): 23–37. (Einstein seems to have been unaware of observations of Brownian motion when he found a theory explaining its characteristics; this paper shows how he realized that some such microscopic motion should exist.)

Smoluchowski, Marian von. "Molekular-kinetische Theorie der Opaleszenz von Gasen im kritischen Zustande, sowie einiger verwandter Erscheinungen." *Annalen der Physik* (Leipzig), 4th vol., 25 (1908): 205–26.

Strutt, Robert John. "Scattering of Light by Dust-free Air, with Artificial Reproduction of the Blue Sky: Preliminary Note." *Proceedings of the Royal Society of London,* ser. A, 94 (1918): 453–59. (While Brücke and Tyndall claimed to have reproduced the blue sky in the laboratory, this feat was only achieved with the fourth Lord Rayleigh's experiments.)

Venkataraman, G. *Raman and His Effect.* Hyderabad: Universities Press, 1995. (An entertaining little book on the Raman effect.)

CHAPTER 9   Ozone's Blue Hour

Adams, C. N., G. N. Plass, and G. W. Kattawar. "The Influence of Ozone and Aerosols on the Brightness and Color of the Twilight Sky." *Journal of the Atmospheric Sciences* 31 (1974): 1662–74.

Anderson, S. M., and K. Mauersberger. "Ozone absorption spectroscopy in search of low-lying electronic states." *Journal of Geophysical Research* D100 (1995): 3033–48.

Dubois, Jean. "Contribution a l'étude de l'ombre de la terre." *Annales de Géophysique* 7 (1951): 103–35. (Jean Dubois was the first to demonstrate that ozone has a visible influence on the colors of the sky; this paper reviews his work of several years of observation.)

Hey, J. D. "From Leonardo to the Graser: Light Scattering in Historical Perspective. Part V." *South African Journal of Science* 82 (1986): 356–60. (Useful as its preceding papers, quoted in Chapters 6 and 7; this paper is focused on the fourth Baron Rayleigh.)

Hulburt, E. O. "Explanation of the Brightness and Color of the Sky, Particularly the Twilight Sky." *Journal of the Optical Society of America* 43 (1953): 113–18. (Hulburt's original paper on the influence of ozone on the colors of twilight.)

Meinel, Aden, and Marjorie Meinel. *Sunsets, Twilights and Evening Skies.* Cambridge: Cambridge University Press, 1983. (An entertaining, informative and beautiful book; unfortunately out of print.)

Rozenberg, V. I. *Twilight: A Study in Atmospheric Optics.* New York: Plenum Press, 1965.

Somerville, Richard C. J. *The Forgiving Air: Understanding Environmental Change.* Berkeley and Los Angeles: University of California Press, 1996.

CHAPTER 10 The Color of Life

Bohren, Craig F. "Multiple scattering and some of its observable consequences." *American Journal of Physics* 55 (1987): 524–33. (Important speculations on how the sky's color depends on the mass of air.)

Bohren, Craig F., and Alistair B. Fraser. "Colors of the Sky." *Physics Teacher* 23 (1985): 267–72.

Budyko, M. I., A. B. Ronov, and A. L. Yanshin. *History of the Earth's Atmosphere.* Berlin: Springer-Verlag, 1987. (Summarizes the path-breaking work done on the subject at the Russian State Hydrological Institute in St. Petersburg; their enthusiasm about the allegedly beneficial effects of global warming has largely subsided.)

Charbonneau, David, Timothy M. Brown, David W. Latham, and Michel Mayor. "Detection of Planetary Transits across a Sun-Like Star." *Astrophysical Journal* 529 (2000): L45–L48.

Charbonneau, David, Timothy M. Brown, Robert W. Noyes, and Ronald L. Gilliland. "Detection of an Extrasolar Planet Atmosphere." *Astrophysical Journal* 568 (2002): 377–84.

Christensen, Gale E. *Greenhouse: The 200-Year History of Global Warming*. New York: Walker and Co., 1999.

Hitchcock, D. R., and J. E. Lovelock. "Life Detection by Atmospheric Analysis." *Icarus* 7 (1967): 149–59.

Houghton, John. *Global Warming: The Complete Briefing*. 3rd ed. Cambridge: Cambridge University Press, 2004. (Probably the best summary of the scientific evidence for global warming, its consequences, and strategies of mitigation.)

Kasting, J. F. "Earth's Early Atmosphere." *Science* 259 (1993): 920–26. (A comparison of Earth's atmosphere with those of the planets Venus and Mars.)

Kasting, James F., and David Catling. "Evolution of a Habitable Planet." *Annual Review of Astronomy and Astrophysics* 41 (2003): 429–63. (Update on Kasting's 1993 paper, places the evolution of Earth in the context of recent studies of extrasolar planets.)

Knoll, Andrew. *Life on a Young Planet: The First Three Billion Years of Evolution on Earth*. Princeton, N.J.: Princeton University Press, 2003. (A gripping tale of the co-evolution of microbial life and Earth's early atmosphere.)

Lovelock, J. *The Ages of Gaia: A Biography of our Living Planet*. Oxford: Oxford University Press, 1988.

Nørretranders, Tor. *Den bla Himmel*. Copenhagen: Munksgaard, 1987. (This Danish book on the color of the sky contains intriguing speculations on its evolution.)

Revelle, Roger, and Hans S. Suess. "Carbon Dioxide Exchanges Between Atmosphere and Ocean and the Question of an Increase of Atmospheric CO_2 During the Past Decades." *Tellus* 9 (1957): 18–27. (This paper has inspired much subsequent research on climate change.)

Sagan, Carl, W. Reid Thompson, Robert Carlson, Donald Gurnett, and Charles Hord. "A Search for Life on Earth with the *Galileo* Spacecraft." *Nature* 365 (1993): 714–21.

Seagar, Sara, E. L. Turner, J. Schafer and E. B. Ford. "Vegetation's Red Edge: A Possible Spectroscopic Biosignature of Extraterrestrial Plants." *Astrobiology* 5 (2005): 372–90.

Smil, Vaclav. *The Earth's Biosphere: Evolution, Dynamics, and Change*. Cambridge, Mass.: MIT Press, 2002. (Contains a wealth of material on the biosphere, from the history of its discovery to the molecular biology of bacteria and the search for life on other planets.)

Turney, Jon. *Lovelock & Gaia: Signs of Life*. Duxford: Icon Books, 2003.

Vernadsky, Vladimir. *The Biosphere*. Translated by David B. Langmuir. New York: Copernicus Press, 1997 [1927]. (First integral English translation of Vernadsky's classic text *Biosfera*; even though it seems that Vernadsky was wrong in almost every detail, his general outlook is valid—and it makes an inspiring reading.)

Vidal-Madjar, A., A. Lecavelier des Etangs, J.-M. Désert, G. E. Ballester, R. Ferlet, G. Hébrard, and M. Mayor. "An Extended Upper Atmosphere Around the Extrasolar Planet HD209458b." *Nature* 422 (2003): 143–46.

Index